Mathematical Principles of Mechanics and Electromagnetism

Part A: Analytical and Continuum Mechanics

MATHEMATICAL CONCEPTS AND METHODS IN SCIENCE AND ENGINEERING

Series Editor: **Angelo Miele**
Mechanical Engineering and Mathematical Sciences
Rice University

Volume 1 **INTRODUCTION TO VECTORS AND TENSORS**
Volume 1: Linear and Multilinear Algebra
Ray M. Bowen and C.-C. Wang

Volume 2 **INTRODUCTION TO VECTORS AND TENSORS**
Volume 2: Vector and Tensor Analysis
Ray M. Bowen and C.-C. Wang

Volume 3 **MULTICRITERIA DECISION MAKING AND DIFFERENTIAL GAMES**
Edited by George Leitmann

Volume 4 **ANALYTICAL DYNAMICS OF DISCRETE SYSTEMS**
Reinhardt M. Rosenberg

Volume 5 **TOPOLOGY AND MAPS**
Taqdir Husain

Volume 6 **REAL AND FUNCTIONAL ANALYSIS**
A. Mukherjea and K. Pothoven

Volume 7 **PRINCIPLES OF OPTIMAL CONTROL THEORY**
R. V. Gamkrelidze

Volume 8 **INTRODUCTION TO THE LAPLACE TRANSFORM**
Peter K. F. Kuhfittig

Volume 9 **MATHEMATICAL LOGIC:** An Introduction to Model Theory
A. H. Lightstone

Volume 11 **INTEGRAL TRANSFORMS IN SCIENCE AND ENGINEERING**
Kurt Bernardo Wolf

Volume 12 **APPLIED MATHEMATICS:** An Intellectual Orientation
Francis J. Murray

Volume 14 **PRINCIPLES AND PROCEDURES OF NUMERICAL ANALYSIS**
Ferenc Szidarovszky and Sidney Yakowitz

Volume 16 **MATHEMATICAL PRINCIPLES OF MECHANICS AND ELECTROMAGNETISM,** Part A: Analytical and Continuum Mechanics
C.-C. Wang

Volume 17 **MATHEMATICAL PRINCIPLES OF MECHANICS AND ELECTROMAGNETISM,** Part B: Electromagnetism and Gravitation
C.-C. Wang

Mathematical Principles of Mechanics and Electromagnetism

Part A: Analytical and Continuum Mechanics

C.-C. Wang
Rice University
Houston, Texas

PLENUM PRESS · NEW YORK AND LONDON

Library of Congress Cataloging in Publication Data

Wang, Chao-Cheng, 1938-
Mathematical principles of mechanics and electromagnetism.

(Mathematical concepts and methods in science and engineering; v. 16–17)
Bibliography: pt. A, p. ; pt, B, p.
Includes Indexes.
CONTENTS: pt. A. Analytical and continuum mechanics.–pt. B. Electromagnetism and gravitation.
1. Mechanics, Analytic. 2. Continuum mechanics. 3. Electromagnetism. 4. Gravitation. I. Title.

QA805.W26 531'.0151 79-11862
ISBN 0-306-40211-4 (v. 16)

A Division of Plenum Publishing Corporation
227 West 17th Street, New York, N.Y. 10011

Printed in the United States of America

To my teacher
CLIFFORD TRUESDELL

Preface to Part A

The purpose of this work is to give an introduction to the mathematical principles of mechanics and of electromagnetism. Part A is concerned with two main subjects in classical mechanics: analytical mechanics and continuum mechanics.

I start in Chapter 1 from Newtonian space–time, which is the basic mathematical model for the event world in all classical theories of physics. On the basis of this model I present the equations of motion for mass points and rigid bodies. Then I derive Lagrange's equations for holonomic systems of mass points and rigid bodies.

My derivation differs in one important aspect from the traditional approach followed by many authors. In the traditional approach a rigid body is regarded as a limiting case of a rigid system of mass points; the number of mass points becomes infinite and the mass of each point becomes infinitesimal in such a way that a finite mass density may be assigned to each part of the body. Lagrange's equations are then derived on the basis of the equations of motion for mass points only, but the results are applied to rigid bodies by using the aforementioned limiting process. Since this limiting process amounts to only a motivation of the equations of motion for a rigid body, I feel that the traditional derivation is not rigorous. In my opinion rigid bodies are primitive concepts like mass points, so that their equations of motion are independent of those for mass points. Hence I derive Lagrange's equations for systems of mass points and rigid bodies from the equations of motion for mass points and for rigid bodies separately. In this way I do not use any argument based on the limiting process in my derivation.

For conservative systems, Lagrange's equations may be transformed into a set of autonomous first-order differential equations, known as the Hamiltonian equations, in phase space. In Chapter 2, I consider the solution operator of the Hamiltonian equations from the standpoint of differential geometry and in the context of a first-order partial differential equation known as the Hamilton–Jacobi equation. The contents of the first two chapters constitute the bulk of general principles in analytical mechanics.

Chapter 3 is devoted to the derivation of the governing equations of motion for deformable bodies. Unlike mass points and rigid bodies, the dynamical responses of which are determined by inertia, deformable bodies have a variety of dynamical responses, which may be described by various constitutive equations. I present just one general class of constitutive equations; bodies characterized by this class are known as simple material bodies.

Constitutive equations for simple material bodies are local models in continuum mechanics, since they are formed by sets of equations of mechanical response for individual body points of the body manifolds. Each distinguished equation of mechanical response characterizes a particular simple material; a simple material body is just a body manifold made up of simple material points. If the points of a body manifold belong to the same simple material, then the body is called materially uniform. The structure of a materially uniform simple material body may be characterized by a single equation of mechanical response and by a distribution of that equation on the body manifold. That distribution may be homogeneous or inhomogeneous.

In Chapter 4, I treat three topics of interest in the theory of simple material bodies in order to illustrate the general principles in continuum mechanics. The first two topics are concerned with homogeneous bodies made up of fluids and isotropic elastic solids, respectively. The third topic is concerned with the geometric structure of inhomogeneous elastic bodies in general.

Throughout this work I have followed a simple, direct, and somewhat old-fashioned approach in order that the text may be followed by advanced undergraduate students with limited background in the elements of the subjects. I set out to present in a clear and rigorous way the basic principles in the two subjects; mathematical generality and elegance are not my primary concern. I hope that this work is helpful to students in grasping the central concepts and results in the subjects. However, I make no claim that the subjects are covered completely in this work. Most of the mathematical preliminaries needed for the formulation of the principles considered in this work may be found in the two-volume work *Introduction to Vectors and Tensors*,* published in this Series (Mathematical Concepts and Methods in Science and Engineering) in 1976.

I am grateful to the Series editor, Angelo Miele, a long-time colleague and a good friend, for permitting me to publish a second work in his series. To Ray Bowen, who has collaborated with me on many other works, I wish to express my thanks for his comments and critical remarks on the preliminary draft.

This work is dedicated to my teacher, Clifford Truesdell, who has directed and guided me on many works, this one included, and who has been a source of inspiration to me for many years. Without his encouragement it would not have been possible for me to undertake and to finish this work.

I take this opportunity to acknowledge also my gratitude to the U.S. National Science Foundation for its support during the preparation of this work.

As always, it is a pleasure to express my appreciation to my wife, Sophia, and to my boys, Ferdie and Ted, for their patience and understanding during the years this work was in progress.

C.-C. Wang

Houston, Texas

* R. M. Bowen and C.-C. Wang, Plenum Press, New York.

Contents of Part A
Analytical and Continuum Mechanics

Contents of Part B
Electromagnetism and Gravitation

1

Lagrangian Mechanics of Particles and Rigid Bodies

In classical mechanics the subject *analytical dynamics* is concerned with motions of particles and rigid bodies. These physical entities may be represented by mathematical models possessing only a finite number of degrees of freedom in their motions. In this chapter we develop the dynamical theory for certain special systems of particles and rigid bodies. The title *Lagrangian mechanics* is chosen because the governing equations of motion for such systems are known as *Lagrange's equations*.

1. Kinematics of Systems of Particles

In classical mechanics the event world is characterized by the *Newtonian space–time* $\mathscr{E}$, which is the union of a family of oriented 3-dimensional Euclidean spaces, $\{\mathscr{E}_\tau, \tau \in \mathscr{T}\}$:

$$\mathscr{E} = \bigcup_{\tau \in \mathscr{T}} \mathscr{E}_\tau, \tag{1.1}$$

where $\mathscr{T}$, the index set of the union, is an oriented 1-dimensional Euclidean space. The index τ is called an *instant*, and the set of instants $\mathscr{T}$ is called the *Newtonian time*. The orientation on $\mathscr{T}$ assigns the set of *past instants* $\{\tau \in \mathscr{T}, \tau < \tau_0\}$ and the set of *future instants* $\{\tau \in \mathscr{T}, \tau > \tau_0\}$ relative to any *present instant* τ_0. At any τ_0 the oriented 3-dimensional Euclidean space $\mathscr{E}_{\tau_0}$ is called the *instantaneous physical space*, and its translation space $\mathscr{V}_{\tau_0}$ is called the *instantaneous translation space*. We do not require

$\mathscr{E}_{\tau_1}$ and $\mathscr{E}_{\tau_2}$ or $\mathscr{V}_{\tau_1}$ and $\mathscr{V}_{\tau_2}$ to be the same for different τ_1 and τ_2. A positive basis in $\mathscr{V}_\tau$ is also said to be *right handed*. In the physical interpretation a right-handed basis is a basis $\{\mathbf{e}_i\}$ which satisfies the following convention: $\mathbf{e}_1$, $\mathbf{e}_2$, and $\mathbf{e}_3$ correspond to the thumb, the index finger, and the middle finger, respectively, of the right hand. The concepts of Euclidean space, the translation space of a Euclidean space, and a positive basis of an oriented vector space are defined in Section 43 of IVT-2.[1]

From (1.1) the Newtonian space–time $\mathscr{E}$ may be represented isometrically by a product space $\mathscr{S}\times\mathscr{R}$, where $\mathscr{S}$ is an oriented 3-dimensional Euclidean space and $\mathscr{R}$ is the set of real numbers. Specifically, we require that $\mathscr{T}$ correspond to $\mathscr{R}$ and that $\mathscr{E}_\tau$ correspond to $\mathscr{S}\times\{t\}$, $\tau\in\mathscr{T}$, by orientation-preserving isometries. The real number t, which corresponds to the instant τ, is called the *time* of the instant, and the oriented 3-dimensional Euclidean space $\mathscr{S}$, which represents the instantaneous spaces, is called the *physical space*. Such a representation of $\mathscr{E}$ is not unique, of course. We call any one such representation a *frame of reference* or an *observer*. We assume that a particular frame of reference has been selected, and we shall now develop the kinematics of systems of particles relative to this particular frame of reference. It is important to remember, however, that the kinematical quantities which we shall define are dependent on this frame.

We consider first the kinematics of a single particle. In Newtonian mechanics a particle is characterized by a positive real number m and by a point $\mathbf{x}(t)\in\mathscr{S}$ at each time $t\in\mathscr{R}$. The number m is called the *mass* or the *inertia* of the particle, and the point $\mathbf{x}(t)$ is called the *position* of the particle at the time t. We can also think of a particle abstractly as just a point p with no intrinsic mathematical structure. Then the position $\mathbf{x}(t)$ may be viewed as the image of a mapping from p to $\mathscr{S}$. In this sense we say that a 1-parameter family, $\{\mathbf{x}(t),\, t\in\mathscr{R}\}$, is a *motion* of p. Thus it is meaningful to consider different motions of the particle p. For simplicity of writing we suppress the argument t and denote the position of p by $\mathbf{x}$. It is understood that in a motion of p, $\mathbf{x}$ is a function of t.

In any motion of p we define the *velocity* $\mathbf{v}$ by

$$\mathbf{v}=\frac{d\mathbf{x}}{dt}, \tag{1.2}$$

[1] Throughout this book the notations IVT-1 and IVT-2 refer to *Introduction to Vectors and Tensors*, Volumes 1 and 2, by R. M. Bowen and C.-C. Wang, in the series Mathematical Concepts and Methods in Science and Engineering, Plenum, New York, 1976.

and the *acceleration* **a** by

$$\mathbf{a} \equiv \frac{d\mathbf{v}}{dt} = \frac{d^2\mathbf{x}}{dt^2}. \tag{1.3}$$

Both **v** and **a** are vectors in the translation space $\mathscr{V}$ of $\mathscr{S}$.

Note. By the remark made before, **v** and **a** depend implicitly on the underlying frame of reference, which is used in the representation of $\mathscr{E}$ by $\mathscr{S} \times \mathscr{R}$.

As explained in Section 44, IVT-2, a rectangular Cartesian coordinate system (x^i, $i = 1, 2, 3$) on $\mathscr{S}$ is defined by a particular point $\mathbf{o} \in \mathscr{S}$, called the *origin*, and by a particular right-handed orthonormal basis $\{\mathbf{e}_i\} \in \mathscr{V}$, called the *coordinate basis* or the *natural basis*, such that

$$\mathbf{x} = \mathbf{o} + x^i\mathbf{e}_i \tag{1.4}$$

or, equivalently,

$$\mathbf{r} \equiv \mathbf{x} - \mathbf{o} = x^i\mathbf{e}_i, \tag{1.5}$$

where **r** is called the *position vector* of **x** relative to **o**. Then the velocity **v** and the acceleration **a** are given by

$$\mathbf{v} = \frac{dx^i}{dt}\mathbf{e}_i, \qquad \mathbf{a} = \frac{dv^i}{dt}\mathbf{e}_i = \frac{d^2x^i}{dt^2}\mathbf{e}_i. \tag{1.6}$$

In other words the components v^i of **v** and a^i of **a** relative to $\{\mathbf{e}_i\}$ are given by

$$v^i = \frac{dx^i}{dt}, \qquad a^i = \frac{dv^i}{dt} = \frac{d^2x^i}{dt^2}. \tag{1.7}$$

From (1.5) we have also

$$\mathbf{v} = \frac{d\mathbf{r}}{dt}, \qquad \mathbf{a} = \frac{d^2\mathbf{r}}{dt^2}. \tag{1.8}$$

Note. In the coordinate representations (1.6) and (1.7) it is important that the origin **o** be fixed in $\mathscr{S}$, and that the basis $\{\mathbf{e}_i\}$ be fixed in $\mathscr{V}$, for all times t. If **o** and $\{\mathbf{e}_i\}$ are time dependent, then the representations become

$$\begin{aligned} \mathbf{v} &= \mathbf{v}^o + x^i\frac{d\mathbf{e}_i}{dt} + \frac{dx^i}{dt}\mathbf{e}_i, \\ \mathbf{a} &= \mathbf{a}^o + 2\frac{dx^i}{dt}\frac{d\mathbf{e}_i}{dt} + x^i\frac{d^2\mathbf{e}_i}{dt^2} + \frac{d^2x^i}{dt^2}\mathbf{e}_i, \end{aligned} \tag{1.9}$$

where $\mathbf{v}^o$ and $\mathbf{a}^o$ denote the velocity and the acceleration of the origin **o**.

In the geometric interpretation a motion of p is just a curve in $\mathscr{S}$, and the velocity is just the tangent vector of the curve as defined in Section 47, IVT-2, the parameter of the curve being the time t. Recall also that in Section 53, IVT-2, we defined the *unit tangent* **s** of a curve by using the arc length parameter s or by normalizing the tangent vector **v**, viz.,

$$\mathbf{s} = \frac{\mathbf{v}}{\|\mathbf{v}\|} = \frac{\mathbf{v}}{v}, \qquad v = \frac{ds}{dt}, \tag{1.10}$$

where $v \equiv \|\mathbf{v}\| \equiv (\mathbf{v} \cdot \mathbf{v})^{1/2}$ is the norm of **v**. Rewriting (1.10) as

$$\mathbf{v} = v\mathbf{s} = \frac{ds}{dt}\,\mathbf{s}, \tag{1.11}$$

we see that the velocity is just a vector which is tangent to the curve described by the motion, and the norm of **v** is just the *speed*, i.e., the rate of change of the arc length s relative to the time t.

Unlike the velocity **v**, the acceleration **a** is generally not tangent to the curve traced out by p in a motion. Indeed, from (1.3) and (1.11), and by using the Serret–Frenet formulas derived in Section 53, IVT-2, we have

$$\mathbf{a} = \frac{dv}{dt}\,\mathbf{s} + v^2\varkappa\mathbf{n} = \frac{d^2s}{dt^2}\,\mathbf{s} + v^2\varkappa\mathbf{n}, \tag{1.12}$$

where $\varkappa$ and **n** denote the curvature and the principal normal of the curve. If $v^2\varkappa \neq 0$, then **a** is not tangent to the curve. Since **a** does not have a component in the direction of the binormal **b** of the curve, it must lie on the osculating plane. We call the first term, $(d^2s/dt^2)\mathbf{s}$, on the right-hand side of (1.12) the *tangential acceleration*, and the second term, $v^2\varkappa\mathbf{n}$, the *normal acceleration* or the *centrifugal acceleration*. The former is a vector tangent to the curve; its norm is the rate of change of the speed, while the latter is a vector normal to the curve on the osculating plane; its norm is equal to v^2/r, where r denotes the radius of curvature; i.e., $r = 1/\varkappa$.

The velocity and the acceleration are the basic kinematical quantities of a motion; other kinematical quantities are derived from them. Specifically, we define the *linear momentum* **l**, the *moment of momentum* **h** (relative to the fixed origin **o**), and the *kinetic energy* e by

$$\mathbf{l} = m\mathbf{v}, \tag{1.13a}$$

$$\mathbf{h} = \mathbf{r}\times\mathbf{l} = m\mathbf{r}\times\mathbf{v}, \tag{1.13b}$$

$$e = \tfrac{1}{2}mv^2. \tag{1.13c}$$

From (1.9) and (1.13) we then have

$$\frac{d\mathbf{l}}{dt} = m\mathbf{a}, \tag{1.14a}$$

$$\frac{d\mathbf{h}}{dt} = m\mathbf{r}\times\mathbf{a}, \tag{1.14b}$$

$$\frac{de}{dt} = m\mathbf{v}\cdot\mathbf{a} = mv\frac{dv}{dt}, \tag{1.14c}$$

where we have used the condition that m is a constant independent of t.

Having defined various kinematical quantities for any motion of a single particle p, we consider next kinematics of a system of particles. Let $\mathscr{P} = \{p^\alpha, \alpha = 1, \dots, N\}$ be a finite set of particles. Then a motion of $\mathscr{P}$ is given by $\{\mathbf{x}^\alpha(t), t \in \mathscr{R}, \alpha = 1, \dots, N\}$. As before we suppress the argument t, and we define the set of velocities $\{\mathbf{v}^\alpha\}$ and the set of accelerations $\{\mathbf{a}^\alpha\}$ by

$$\mathbf{v}^\alpha = \frac{d\mathbf{x}^\alpha}{dt}, \qquad \mathbf{a}^\alpha = \frac{d\mathbf{v}^\alpha}{dt} = \frac{d^2\mathbf{x}^\alpha}{dt^2}, \qquad \alpha = 1, \dots, N. \tag{1.15}$$

An important kinematical concept for a system of particles is the *center of mass*. At any time t in a motion of $\mathscr{P}$ the center of mass is the point $\mathbf{x}^c = \mathbf{x}^c(t)$ whose position vector $\mathbf{r}^c = \mathbf{x}^c - \mathbf{o}$ is given by

$$\mathbf{r}^c = \frac{\sum_{\alpha=1}^{N} m^\alpha \mathbf{r}^\alpha}{\sum_{\alpha=1}^{N} m^\alpha} = \frac{1}{M}\sum_{\alpha=1}^{N} m^\alpha \mathbf{r}^\alpha, \qquad M = \sum_{\alpha=1}^{N} m^\alpha, \tag{1.16}$$

where m^α and $\mathbf{r}^\alpha$ denote the mass and the position vector $\mathbf{r}^\alpha = \mathbf{x}^\alpha - \mathbf{o}$ of p^α, respectively. It can be shown easily that $\mathbf{x}^c$ is determined uniquely by the set of positions $\{\mathbf{x}^\alpha\}$ of $\mathscr{P}$, independent of the choice of the origin $\mathbf{o}$. Consequently we can regard $\mathbf{x}^c$ as the position occupied by a fictitious particle p^c. As we shall see, it is convenient to regard the mass of p^c as being the total mass M.

Indeed, from (1.16) the velocity $\mathbf{v}^c$ and the acceleration $\mathbf{a}^c$ of p^c are given by

$$\mathbf{v}^c = \frac{1}{M}\sum_{\alpha=1}^{N} m^\alpha \mathbf{v}^\alpha, \qquad \mathbf{a}^c = \frac{1}{M}\sum_{\alpha=1}^{N} m^\alpha \mathbf{a}^\alpha. \tag{1.17}$$

As a result, the total linear momentum $\mathbf{L}$ of $\mathscr{P}$ is given by

$$\mathbf{L} = \sum_{\alpha=1}^{N} m^\alpha \mathbf{v}^\alpha = M\mathbf{v}^c, \tag{1.18}$$

and its rate of change is given by

$$\frac{d\mathbf{L}}{dt} = \sum_{\alpha=1}^{N} m^{\alpha}\mathbf{a}^{\alpha} = M\mathbf{a}^{c}. \tag{1.19}$$

Hence if we take the mass of p^c to be the total mass M, then the total linear momentum of the system is the same as that of p^c.

Representations for the total moment of momentum and the total kinetic energy are somewhat more complex than the representation for the total linear momentum. First, we define the position vector $\mathbf{r}^{\alpha,c}$, the velocity $\mathbf{v}^{\alpha,c}$, and the acceleration $\mathbf{a}^{\alpha,c}$ of p^α relative to p^c by

$$\mathbf{r}^{\alpha,c} = \mathbf{r}^{\alpha} - \mathbf{r}^{c} = \mathbf{x}^{\alpha} - \mathbf{x}^{c}, \qquad \mathbf{v}^{\alpha,c} = \mathbf{v}^{\alpha} - \mathbf{v}^{c}, \qquad \mathbf{a}^{\alpha,c} = \mathbf{a}^{\alpha} - \mathbf{a}^{c}. \tag{1.20}$$

From (1.17) the mass-weighted averages of $\mathbf{r}^{\alpha,c}$, $\mathbf{v}^{\alpha,c}$, $\mathbf{a}^{\alpha,c}$ all vanish:

$$\sum_{\alpha=1}^{N} m^{\alpha}\mathbf{r}^{\alpha,c} = \sum_{\alpha=1}^{N} m^{\alpha}\mathbf{v}^{\alpha,c} = \sum_{\alpha=1}^{N} m^{\alpha}\mathbf{a}^{\alpha,c} = \mathbf{0}. \tag{1.21}$$

Using these identities, we can express the total moment of momentum $\mathbf{H}$ of $\mathscr{P}$ relative to $\mathbf{o}$ by

$$\mathbf{H} = \sum_{\alpha=1}^{N} m^{\alpha}\mathbf{r}^{\alpha}\times\mathbf{v}^{\alpha} = M\mathbf{r}^{c}\times\mathbf{v}^{c} + \sum_{\alpha=1}^{N} m^{\alpha}\mathbf{r}^{\alpha,c}\times\mathbf{v}^{\alpha,c}. \tag{1.22}$$

The leading term on the right-hand side of (1.22) is, of course, just the moment of momentum of p^c with mass M. The next term can be identified as the total moment of momentum of $\mathscr{P}$ relative to p^c. Thus the total moment of momentum of the system is the sum of two parts: the first part being the moment of momentum of the center of mass and the second part being the moment of momentum relative to the center of mass. Direct differentiation of (1.22) with respect to time yields

$$\frac{d\mathbf{H}}{dt} = M\mathbf{r}^{c}\times\mathbf{a}^{c} + \sum_{\alpha=1}^{N} m^{\alpha}\mathbf{r}^{\alpha,c}\times\mathbf{a}^{\alpha,c}. \tag{1.23}$$

Representation for the total kinetic energy E of the system $\mathscr{P}$ may be derived in a similar way, and the result is

$$E = \frac{1}{2}\sum_{\alpha=1}^{N} m^{\alpha}\mathbf{v}^{\alpha}\cdot\mathbf{v}^{\alpha} = \frac{1}{2}M\mathbf{v}^{c}\cdot\mathbf{v}^{c} + \frac{1}{2}\sum_{\alpha=1}^{N} m^{\alpha}\mathbf{v}^{\alpha,c}\cdot\mathbf{v}^{\alpha,c}, \tag{1.24}$$

which means that E is the sum of the kinetic energy of the center of mass and the kinetic energy relative to the center of mass. Differentiation of

(1.24) yields the rate of change of E:

$$\frac{dE}{dt} = M\mathbf{v}^c \cdot \mathbf{a}^c + \sum_{\alpha=1}^{N} m^\alpha \mathbf{v}^{\alpha,c} \cdot \mathbf{a}^{\alpha,c}. \tag{1.25}$$

So far, we have considered various kinematical quantities for a particle or for a system of particles which are allowed to move arbitrarily in the Euclidean space $\mathscr{S}$. Physically, of course, particles do not move arbitrarily. Indeed, each motion of a particle is governed by certain laws of dynamics, which we shall develop later. Not only that, motions of particles may be constrained in a certain predetermined way; e.g., a system $\mathscr{P}$ may be constrained in such a way that the distances

$$d(\mathbf{x}^\alpha, \mathbf{x}^\beta) = \| \mathbf{x}^\alpha - \mathbf{x}^\beta \| \tag{1.26}$$

remain fixed independent of time for all pairs p^α and p^β in $\mathscr{P}$. If $\mathscr{P}$ is subject to such a constraint, then it is called a *rigid system*. For a constrained system in general, since the positions $\mathbf{x}^\alpha$ may not be arbitrary, it is more convenient to express a motion in terms of certain *generalized coordinates*, which characterize the constraint and are allowed to be arbitrary. We shall discuss the kinematics of constrained systems of particles as well as rigid bodies in Section 3.

2. Kinematics of a Rigid Body

Unlike the mathematical model for a particle, which is a point p assigned with a positive mass m, a rigid body is characterized by a domain $\mathscr{B}$ in an oriented 3-dimensional Euclidean space and by a finite positive mass measure μ on $\mathscr{B}$. The total mass measure of $\mathscr{B}$, viz.,

$$M = \mu(\mathscr{B}) = \int_{\mathscr{B}} d\mu, \tag{2.1}$$

is called the *mass* of $\mathscr{B}$. A motion of $\mathscr{B}$ is defined by a 1-parameter family of orientation-preserving isometries of $\mathscr{B}$ into $\mathscr{S}$, viz.,

$$\mathbf{x}(t, \cdot) : \mathscr{B} \to \mathscr{S}, \qquad t \in \mathscr{R}. \tag{2.2}$$

We call $\mathscr{B}$ a *rigid body*, since for an isometry the distances

$$d(p, q) = \| \mathbf{x}(t, p) - \mathbf{x}(t, q) \| \tag{2.3}$$

remain constant for all pairs of points $p, q \in \mathscr{B}$ and for all times t.

We define the center of mass p^c of $\mathscr{B}$ by requiring that

$$\int_{\mathscr{B}} \mathbf{r}^{p,c}\, d\mu = \mathbf{0}, \tag{2.4}$$

where $\mathbf{r}^{p,c}$ denotes the position vector of a point $p \in \mathscr{B}$ relative to p^c, viz.,

$$\mathbf{r}^{p,c} = p - p^c. \tag{2.5}$$

Note. The condition (2.4) is equivalent to the condition (1.21a) except that a rigid body generally contains infinitely many points, so the summation in (1.21a) is replaced by the integration in (2.4). If we choose an arbitrary origin o in the Euclidean manifold of $\mathscr{B}$, then the position vector $\mathbf{r}^c$ of p^c relative to o is given by

$$\mathbf{r}^c = \frac{1}{M} \int_{\mathscr{B}} \mathbf{r}^p\, d\mu, \tag{2.6}$$

where $\mathbf{r}^p$ denotes the position vector of the point of integration p relative to o; i.e.,

$$\mathbf{r}^p = p - o. \tag{2.7}$$

To characterize a point $p \in \mathscr{B}$, we may use a rectangular Cartesian coordinate system (y^i) on $\mathscr{B}$. We choose the origin of the coordinate system at the center of mass p^c, and we call the natural basis $\{\mathbf{f}_i\}$ of (y^i) an *imbedded basis* at p^c. Then (2.4) has the coordinate form

$$\int_{\mathscr{B}} y^i\, d\mu = 0, \qquad i = 1, 2, 3, \tag{2.8}$$

since the components of $\mathbf{r}^{p,c}$ relative to $\{\mathbf{f}_i\}$ are just the imbedded coordinates (y^i) of p; i.e.,

$$\mathbf{r}^{p,c} = y^i \mathbf{f}_i. \tag{2.9}$$

In a motion $\{\mathbf{x}(t, \cdot), t \in \mathscr{R}\}$ of $\mathscr{B}$ the mappings $\mathbf{x}(t, \cdot)$ are isometries of $\mathscr{B}$ into $\mathscr{S}$. Consequently the imbedded coordinate system (y^i) is transformed into a rectangular Cartesian coordinate system on the image $\mathbf{x}(t, \mathscr{B}) \subset \mathscr{S}$ for each $t \in \mathscr{R}$. Not only that, the origin of (y^i) remains the center of mass of $\mathscr{B}$ at time t, since the condition (2.8) still holds, and the integrand y^i is still the component of the position vector $\mathbf{r}^p$ relative to the origin. Furthermore, the imbedded basis is mapped onto an orthonormal basis, which we shall denote by $\{\mathbf{f}_i\}$ also, for the physical translation space

$\mathscr{V}$. It is important to note, however, that the origin $\mathbf{x}^c = \mathbf{x}(t, p^c)$ in $\mathscr{S}$ and the natural basis $\{\mathbf{f}_i\}$ of (y^i) in $\mathscr{V}$ both depend on time. As a result when we calculate the velocity and the acceleration of points in $\mathscr{B}$ in terms of (y^i), we must use the general formulas (1.9).

Specifically, for any $p \in \mathscr{B}$ with imbedded coordinates (y^i), the position $\mathbf{x}^p = \mathbf{x}(t, p)$ and the velocity $\mathbf{v}^p = \mathbf{v}(t, p)$ are given by

$$\mathbf{x}^p = \mathbf{x}^c + \mathbf{r}^{p,c} = \mathbf{x}^c + y^i\mathbf{f}_i, \qquad \mathbf{v}^p = \mathbf{v}^c + y^i\frac{d\mathbf{f}_i}{dt}. \tag{2.10}$$

Here we have used the fact that for a particular point $p \in \mathscr{B}$ the imbedded coordinates (y^i) are constants independent of t. By using the fact that the imbedded basis $\{\mathbf{f}_i\}$ is an orthonormal basis in $\mathscr{V}$, we can express the time rate $d\mathbf{f}_i/dt$ by

$$\frac{d\mathbf{f}_i}{dt} = \Omega_{ij}\mathbf{f}_j, \tag{2.11}$$

where $[\Omega_{ij}]$ is a skew-symmetric matrix, since

$$\Omega_{ij} = \mathbf{f}_j \cdot \frac{d\mathbf{f}_i}{dt} = \frac{d}{dt}(\mathbf{f}_j \cdot \mathbf{f}_i) - \mathbf{f}_i \cdot \frac{d\mathbf{f}_j}{dt} = \frac{d}{dt}(\delta_{ji}) - \Omega_{ji} = -\Omega_{ji}. \tag{2.12}$$

Consequently we can rewrite (2.11) as

$$\frac{d\mathbf{f}_i}{dt} = \boldsymbol{\omega} \times \mathbf{f}_i, \tag{2.13}$$

where $\boldsymbol{\omega}$ is given by

$$\boldsymbol{\omega} = \Omega_{23}\mathbf{f}_1 - \Omega_{13}\mathbf{f}_2 + \Omega_{12}\mathbf{f}_3 = \tfrac{1}{2}\varepsilon_{ijk}\Omega_{jk}\mathbf{f}_i. \tag{2.14}$$

Substituting (2.13) into (2.10), we obtain

$$\mathbf{v}^p = \mathbf{v}^c + \boldsymbol{\omega} \times y^i\mathbf{f}_i = \mathbf{v}^c + \boldsymbol{\omega} \times \mathbf{r}^{p,c}. \tag{2.15}$$

This is the *velocity representation formula* relative to the center of mass. The vector $\boldsymbol{\omega}$, which is given explicitly by (2.14), is called the *angular velocity* of the rigid body.

We can generalize the velocity representation formula to an arbitrary imbedded origin $q \in \mathscr{B}$. Indeed, from (2.15) for the point q, we have

$$\mathbf{v}^q = \mathbf{v}^c + \boldsymbol{\omega} \times \mathbf{r}^{q,c}. \tag{2.16}$$

Subtracting (2.16) from (2.15), we get

$$\mathbf{v}^p - \mathbf{v}^q = \boldsymbol{\omega}\times(\mathbf{r}^{p,c} - \mathbf{r}^{q,c}) = \boldsymbol{\omega}\times\mathbf{r}^{p,q}, \tag{2.17}$$

where $\mathbf{r}^{p,q}$ is the position vector of $\mathbf{x}(t, p)$ relative to $\mathbf{x}(t, q)$; i.e.,

$$\mathbf{r}^{p,q} = \mathbf{x}(t, p) - \mathbf{x}(t, q). \tag{2.18}$$

Transferring the term $\mathbf{v}^q$ in (2.17) to the right-hand side, we obtain

$$\mathbf{v}^p = \mathbf{v}^q + \boldsymbol{\omega}\times\mathbf{r}^{p,q}. \tag{2.19}$$

This is the *velocity representation formula* relative to the arbitrary point $q \in \mathscr{B}$. From (2.19) we see that the angular velocity $\boldsymbol{\omega}$ is independent of the imbedded origin used in the velocity representation.

From the preceding analysis we see that a motion of a rigid body $\mathscr{B}$ may be characterized by the list $\{\mathbf{x}^c, \mathbf{f}_i, i = 1, 2, 3\}$, which can be viewed as a moving frame in space. At any time t, $\mathbf{x}^c$ is the position of the center of mass of $\mathscr{B}$ in $\mathbf{x}(t, \mathscr{B})$, and $\{\mathbf{f}_i\}$ is the position of the imbedded basis attached to $\mathbf{x}^c$. Given the list $\{\mathbf{x}^c, \mathbf{f}_i, i = 1, 2, 3\}$ we can determine the position of any point $p \in \mathscr{B}$ with imbedded coordinates (y^i) by (2.10) and, then, we can calculate the velocity of p by the formula (2.15).

Now to characterize the list $\{\mathbf{x}^c, \mathbf{f}_i\}$ we may refer to a fixed rectangular Cartesian coordinate system (x^i) with origin $\mathbf{o}$ and natural basis $\{\mathbf{e}_i\}$ in $\mathscr{S}$. Specifically, $\mathbf{x}^c$ is just the position of a moving point, so it can be characterized by the coordinates (ξ^i), viz.,

$$\mathbf{x}^c = \mathbf{o} + \xi^i\mathbf{e}_i, \tag{2.20a}$$

$$\mathbf{v}^c = \frac{d\xi^i}{dt}\mathbf{e}_i, \tag{2.20b}$$

as in the preceding section. As far as the moving basis $\{\mathbf{f}_i\}$ is concerned, we can use the component representation

$$\mathbf{f}_i = Q_{ij}\mathbf{e}_j, \tag{2.21}$$

where $[Q_{ij}]$ is a rotation matrix. The collection of all rotation matrices form a continuous group $\mathscr{SO}(3)$. As explained in Section 63, IVT-2, $\mathscr{SO}(3)$ is a 3-dimensional compact manifold in the 9-dimensional Euclidean space $\mathscr{R}^9$ of 3×3 matrices. In particular, it takes three coordinates to characterize a particular $[Q_{ij}]$ in $\mathscr{SO}(3)$. Of course, there are many coordinate systems on the manifold $\mathscr{SO}(3)$.

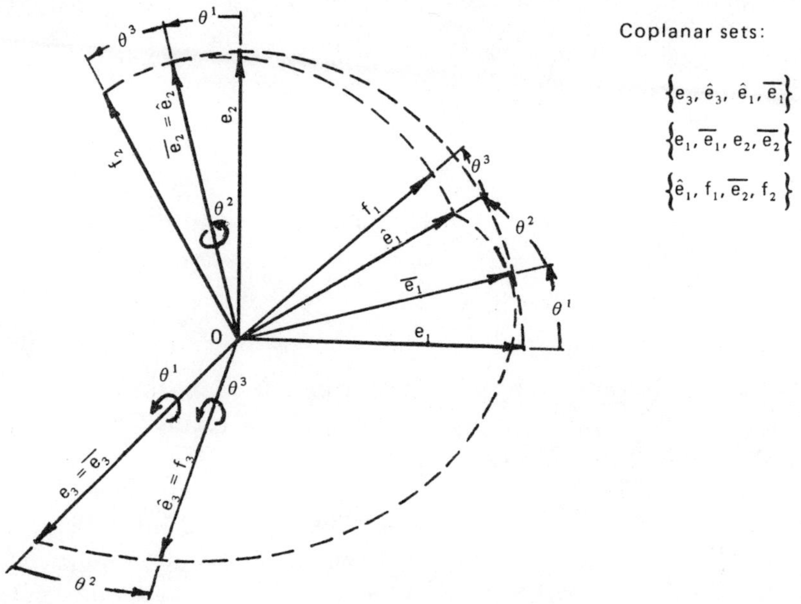

Figure 1.

One coordinate system, called the *Eulerian angles* (θ^i, $i = 1, 2, 3$), is defined as indicated in Fig. 1. Specifically, $\{\bar{\mathbf{e}}_i\}$ is obtained from $\{\mathbf{e}_i\}$ by a rotation of θ^1 about the axis $\mathbf{e}_3$, viz.,

$$\begin{pmatrix} \bar{\mathbf{e}}_1 \\ \bar{\mathbf{e}}_2 \\ \bar{\mathbf{e}}_3 \end{pmatrix} = \begin{pmatrix} \cos\theta^1 & \sin\theta^1 & 0 \\ -\sin\theta^1 & \cos\theta^1 & 0 \\ 0 & 0 & 1 \end{pmatrix} \begin{pmatrix} \mathbf{e}_1 \\ \mathbf{e}_2 \\ \mathbf{e}_3 \end{pmatrix}. \tag{2.22}$$

Similarly, $\{\hat{\mathbf{e}}_i\}$ is obtained from $\{\bar{\mathbf{e}}_i\}$ by a rotation of θ^2 about the axis $\bar{\mathbf{e}}_2$, viz.,

$$\begin{pmatrix} \hat{\mathbf{e}}_1 \\ \hat{\mathbf{e}}_2 \\ \hat{\mathbf{e}}_3 \end{pmatrix} = \begin{pmatrix} \cos\theta^2 & 0 & -\sin\theta^2 \\ 0 & 1 & 0 \\ \sin\theta^2 & 0 & \cos\theta^2 \end{pmatrix} \begin{pmatrix} \bar{\mathbf{e}}_1 \\ \bar{\mathbf{e}}_2 \\ \bar{\mathbf{e}}_3 \end{pmatrix}. \tag{2.23}$$

Finally, $\{\mathbf{f}_i\}$ is obtained from $\{\hat{\mathbf{e}}_i\}$ by a rotation of θ^3 about the axis $\hat{\mathbf{e}}_3$, viz.,

$$\begin{pmatrix} \mathbf{f}_1 \\ \mathbf{f}_2 \\ \mathbf{f}_3 \end{pmatrix} = \begin{pmatrix} \cos\theta^3 & \sin\theta^3 & 0 \\ -\sin\theta^3 & \cos\theta^3 & 0 \\ 0 & 0 & 1 \end{pmatrix} \begin{pmatrix} \hat{\mathbf{e}}_1 \\ \hat{\mathbf{e}}_2 \\ \hat{\mathbf{e}}_3 \end{pmatrix}. \tag{2.24}$$

Combining (2.22)–(2.24), we see that the bases $\{\mathbf{e}_i\}$ and $\{\mathbf{f}_i\}$ are related by

$$\begin{pmatrix}\mathbf{f}_1\\ \mathbf{f}_2\\ \mathbf{f}_3\end{pmatrix} = \begin{pmatrix}\cos\theta^3 & \sin\theta^3 & 0\\ -\sin\theta^3 & \cos\theta^3 & 0\\ 0 & 0 & 1\end{pmatrix}\begin{pmatrix}\cos\theta^2 & 0 & -\sin\theta^2\\ 0 & 1 & 0\\ \sin\theta^2 & 0 & \cos\theta^2\end{pmatrix}\begin{pmatrix}\cos\theta^1 & \sin\theta^1 & 0\\ -\sin\theta^1 & \cos\theta^1 & 0\\ 0 & 0 & 1\end{pmatrix}\begin{pmatrix}\mathbf{e}_1\\ \mathbf{e}_2\\ \mathbf{e}_3\end{pmatrix}, \tag{2.25a}$$

or:

$$\begin{pmatrix}\mathbf{f}_1\\ \mathbf{f}_2\\ \mathbf{f}_3\end{pmatrix} = \begin{pmatrix}\cos\theta^1\cos\theta^2\cos\theta^3 - \sin\theta^1\sin\theta^3 & \sin\theta^1\cos\theta^2\cos\theta^3 + \cos\theta^1\sin\theta^3 & -\sin\theta^2\cos\theta^3\\ -\cos\theta^1\cos\theta^2\sin\theta^3 - \sin\theta^1\cos\theta^3 & -\sin\theta^1\cos\theta^2\sin\theta^3 + \cos\theta^1\cos\theta^3 & \sin\theta^2\sin\theta^3\\ \cos\theta^1\sin\theta^2 & \sin\theta^1\sin\theta^2 & \cos\theta^2\end{pmatrix}\begin{pmatrix}\mathbf{e}_1\\ \mathbf{e}_2\\ \mathbf{e}_3\end{pmatrix}, \tag{2.25b}$$

where the coefficient matrix on the right-hand side is just the matrix $[Q_{ij}]$ in (2.21). Thus (2.25) gives a representation of a rotation matrix $[Q_{ij}]$ in general in terms of the Eulerian angles (θ^i).

We can obtain the angular velocity $\boldsymbol{\omega}$ by using the formulas (2.11), (2.14), and (2.25). It is more convenient to differentiate (2.25a) with respect to t first by the product rule, and then changing the basis from $\{\mathbf{e}_i\}$ to $\{\mathbf{f}_i\}$ by the inverse of (2.25a). Notice that for an orthogonal matrix the inverse is just the transpose. The final result is

$$\begin{aligned}\boldsymbol{\omega} &= \omega^i\mathbf{f}_i, && (2.26a)\\ &= \omega^1\mathbf{f}_1 + \omega^2\mathbf{f}_2 + \omega^3\mathbf{f}_3, && (2.26b)\\ &= (-\dot\theta^1\sin\theta^2\cos\theta^3 + \dot\theta^2\sin\theta^3)\mathbf{f}_1 + (\dot\theta^1\sin\theta^2\sin\theta^3 + \dot\theta^2\cos\theta^3)\mathbf{f}_2\\ &\quad + (\dot\theta^1\cos\theta^2 + \dot\theta^3)\mathbf{f}_3, && (2.26c)\\ &= \dot\theta^1(-\sin\theta^2\cos\theta^3\mathbf{f}_1 + \sin\theta^2\sin\theta^3\mathbf{f}_2 + \cos\theta^2\mathbf{f}_3)\\ &\quad + \dot\theta^2(\sin\theta^3\mathbf{f}_1 + \cos\theta^3\mathbf{f}_2) + \dot\theta^3\mathbf{f}_3, && (2.26d)\\ &= \dot\theta^1\mathbf{e}_3 + \dot\theta^2\bar{\mathbf{e}}_2 + \dot\theta^3\hat{\mathbf{e}}_3. && (2.26e)\end{aligned}$$

Note. The last line above, (2.26e), follows directly from the differentiation of the three matrices in (2.25a) by using the product rule. In the physical interpretation (2.25a) is called the *multiplicative decomposition* of a finite rotation, while (2.26e) is called the *additive decomposition* of an infinitesimal rotation.

We can rewrite the component form (2.26c) as

$$\begin{pmatrix}\omega^1\\ \omega^2\\ \omega^3\end{pmatrix} = \begin{pmatrix}-\sin\theta^2\cos\theta^3 & \sin\theta^3 & 0\\ \sin\theta^2\sin\theta^3 & \cos\theta^3 & 0\\ \cos\theta^2 & 0 & 1\end{pmatrix}\begin{pmatrix}\dot\theta^1\\ \dot\theta^2\\ \dot\theta^3\end{pmatrix} \tag{2.27}$$

or, more compactly,

$$\omega^i = W^{ij}\dot{\theta}^j, \tag{2.28}$$

where $[W^{ij}]$ denotes the coefficient matrix on the right-hand side of (2.27). The matrix $[W^{ij}]$ is not orthogonal; it is nonsingular when θ^2 is not an integral multiple of π.

Summarizing the results, we see that a configuration of $\mathscr{B}$ (i.e., an orientation-preserving isometry of $\mathscr{B}$ into $\mathscr{S}$) can be characterized by six parameters ($\xi^i, \theta^i, i = 1, 2, 3$), which determine uniquely the position of the moving imbedded frame $\{\mathbf{x}^c, \mathbf{f}_i, i = 1, 2, 3\}$ relative to a fixed frame $\{\mathbf{o}, \mathbf{e}_i, i = 1, 2, 3\}$ in space. In any motion of $\mathscr{B}$ the parameters (ξ^i, θ^i, $i = 1, 2, 3$) are generally functions of time. From these functions we can calculate the angular velocity $\boldsymbol{\omega}$ by (2.26) or (2.27), and then the velocity of an arbitrary point $p \in \mathscr{B}$ by (2.15), with $\mathbf{v}^c$ given by (2.20b). As before the velocity is the basic kinematical quantity, from which we can derive various other kinematical quantities.

Specifically, the linear momentum $\mathbf{L}$ is defined by

$$\mathbf{L} \equiv \int_{\mathscr{B}} \mathbf{v}^p \, d\mu = \int_{\mathscr{B}} (\mathbf{v}^c + \boldsymbol{\omega} \times \mathbf{r}^{p,c}) \, d\mu = M\mathbf{v}^c, \tag{2.29}$$

where we have used the basic condition (2.4) for the center of mass. The result (2.29) is similar to the formula (1.18), except that $\mathscr{B}$ is generally formed by infinitely many points. As before the rate of change of the linear momentum is given by

$$\frac{d\mathbf{L}}{dt} = \int_{\mathscr{B}} \mathbf{a}^p \, d\mu = M\mathbf{a}^c. \tag{2.30}$$

The moment of momentum $\mathbf{H}$ relative to the fixed origin $\mathbf{o}$ in space is defined by

$$\begin{aligned} \mathbf{H} &\equiv \int_{\mathscr{B}} \mathbf{r}^p \times \mathbf{v}^p \, d\mu = \int_{\mathscr{B}} \mathbf{r}^p \times (\mathbf{v}^c + \boldsymbol{\omega} \times \mathbf{r}^{p,c}) \, d\mu \\ &= M\mathbf{r}^c \times \mathbf{v}^c + \int_{\mathscr{B}} \mathbf{r}^{p,c} \times \boldsymbol{\omega} \times \mathbf{r}^{p,c} \, d\mu, \end{aligned} \tag{2.31}$$

where we have used again the basic condition (2.4). The formula (2.31) is similar to the formula (1.22). Indeed, the first term is just the moment of momentum of the center of mass assigned with the mass M, while the second term is the moment of momentum relative to the center of mass. This second term can best be calculated by using the imbedded coordinate system,

$$\int_{\mathscr{B}} \mathbf{r}^{p,c} \times \boldsymbol{\omega} \times \mathbf{r}^{p,c}\, d\mu = \left[\int_{\mathscr{B}} (y^k y^k \delta^{ij} - y^i y^j)\, d\mu\right] \omega^j \mathbf{f}_i \equiv I^{ij} \omega^j \mathbf{f}_i, \tag{2.32}$$

where ω^i denote the components of $\boldsymbol{\omega}$ relative to $\{\mathbf{f}_i\}$ as shown in (2.26).

Since the imbedded coordinates (y^i) are the components of the position vector relative to the center of mass, from (2.32) we see clearly that the matrix $[I^{ij}]$ is the component matrix of a certain second-order tensor[2] $\mathbf{I}$ over the translation space of the Euclidean manifold of $\mathscr{B}$. Specifically, $\mathbf{I}$ is given by

$$\mathbf{I} = \int_{\mathscr{B}} (\| \mathbf{r}^{p,c} \|^2 \mathbf{1} - \mathbf{r}^{p,c} \otimes \mathbf{r}^{p,c})\, d\mu, \tag{2.33}$$

where $\mathbf{1}$ denotes the identity tensor.[2] We call $\mathbf{I}$ the *inertia tensor* of $\mathscr{B}$ relative to the center of mass p^c. It is easily seen that $\mathbf{I}$ is a symmetric tensor. From the Schwarz inequality (cf. Section 12, IVT-1) and from the assumption that μ is a positive measure, we can show that $\mathbf{I}$ is positive definite. Indeed, the quadratic form of $\mathbf{I}$ is given by

$$\mathbf{u} \cdot \mathbf{I}\mathbf{u} = \int_{\mathscr{B}} [\| \mathbf{r}^{p,c} \|^2 \| \mathbf{u} \|^2 - (\mathbf{r}^{p,c} \cdot \mathbf{u})^2]\, d\mu > 0 \tag{2.34}$$

for all nonvanishing vectors $\mathbf{u}$ in the translation space of the Euclidean manifold of $\mathscr{B}$. Using the inertia tensor $\mathbf{I}$, we can rewrite the formula (2.31) as

$$\mathbf{H} = M\mathbf{r}^c \times \mathbf{v}^c + \mathbf{I}\boldsymbol{\omega}, \tag{2.35}$$

where the component form of $\mathbf{I}\boldsymbol{\omega}$ is given by (2.32).

We can calculate the rate of change of $\mathbf{H}$ directly from (2.35),

$$\frac{d\mathbf{H}}{dt} = M\mathbf{r}^c \times \mathbf{a}^c + \frac{d}{dt}(\mathbf{I}\boldsymbol{\omega}), \tag{2.36}$$

where the second term on the right-hand side can be derived in the following way: From the component form (2.32) and the formula (2.13) we get

$$\frac{d}{dt}(\mathbf{I}\boldsymbol{\omega}) = I^{ij} \frac{d\omega^j}{dt} \mathbf{f}_i + I^{ij}\omega^j \frac{d\mathbf{f}_i}{dt} \tag{2.37a}$$

$$= I^{ij} \frac{d\omega^j}{dt} \mathbf{f}_i + \boldsymbol{\omega} \times I^{ij}\omega^j \mathbf{f}_i \tag{2.37b}$$

$$= \mathbf{I}\dot{\boldsymbol{\omega}} + \boldsymbol{\omega} \times \mathbf{I}\boldsymbol{\omega}, \tag{2.37c}$$

[2] In this chapter the symbol $\mathbf{I}$ denotes the inertia tensor, while the symbol $\mathbf{1}$ denotes the identity tensor. Elsewhere, $\mathbf{I}$ denotes the identity tensor.

where in the last equation, (2.37c), we have used the fact that

$$\dot{\boldsymbol{\omega}} = \frac{d\boldsymbol{\omega}}{dt} = \frac{d\omega^i}{dt}\mathbf{f}_i + \omega^i\frac{d\mathbf{f}_i}{dt} = \frac{d\omega^i}{dt}\mathbf{f}_i + \boldsymbol{\omega}\times\omega^i\mathbf{f}_i$$
$$= \frac{d\omega^i}{dt}\mathbf{f}_i + \boldsymbol{\omega}\times\boldsymbol{\omega} = \frac{d\omega^i}{dt}\mathbf{f}_i. \tag{2.38}$$

Substituting (2.37) into (2.36), we obtain

$$\frac{d\mathbf{H}}{dt} = M\mathbf{r}^c\times\mathbf{a}^c + \mathbf{I}\dot{\boldsymbol{\omega}} + \boldsymbol{\omega}\times\mathbf{I}\boldsymbol{\omega}. \tag{2.39}$$

Next, the kinetic energy E is defined by

$$E = \int_{\mathscr{B}} \frac{1}{2}\,\|\,\mathbf{v}^p\,\|^2\,d\mu = \int_{\mathscr{B}} \frac{1}{2}(\mathbf{v}^c + \boldsymbol{\omega}\times\mathbf{r}^{p,c})\cdot(\mathbf{v}^c + \boldsymbol{\omega}\times\mathbf{r}^{p,c})\,d\mu$$
$$= \frac{1}{2}M\,\|\,\mathbf{v}^c\,\|^2 + \frac{1}{2}\boldsymbol{\omega}\cdot\mathbf{I}\boldsymbol{\omega}, \tag{2.40}$$

where we have used (2.4) and (2.33). This formula is the counterpart of the formula (1.24). From (2.40) we can calculate the rate of change of E by

$$\frac{dE}{dt} = M\mathbf{v}^c\cdot\mathbf{a}^c + \frac{1}{2}\dot{\boldsymbol{\omega}}\cdot\mathbf{I}\boldsymbol{\omega} + \frac{1}{2}\boldsymbol{\omega}\cdot\frac{d}{dt}(\mathbf{I}\boldsymbol{\omega}) = M\mathbf{v}^c\cdot\mathbf{a}^c + \boldsymbol{\omega}\cdot\mathbf{I}\dot{\boldsymbol{\omega}}, \tag{2.41}$$

where we have used (2.37) and the symmetry of $\mathbf{I}$.

As remarked in the preceding section, the kinematics is developed here for a rigid body which is free of any constraints, so the six parameters $(\xi^i, \theta^i, i = 1, 2, 3)$ are independent and arbitrary. If certain constraints are imposed, then the parameters are no longer independent. In that case we may replace the parameters by a set of generalized coordinates, which are independent and are allowed to be arbitrary. We shall develop the kinematics of constrained systems of particles and rigid bodies in the following section.

3. Kinematics of Holonomic Systems of Particles and Rigid Bodies

We now consider a system

$$\mathscr{Q} = \{p^\alpha, \mathscr{B}^\beta, \alpha = 1, \ldots, N, \beta = 1, \ldots, K\} \tag{3.1}$$

consisting in N particles and K rigid bodies. To characterize a configuration of $\mathscr{Q}$, we use N lists of coordinates $(x^{\alpha,i})$ for the N particles and K lists of parameters $(\xi^{\beta,i}, \theta^{\beta,i})$ for the K rigid bodies, all referred to a fixed rectangular Cartesian coordinate system (x^i) on $\mathscr{E}$, defined by the origin $\mathbf{o}$ and the coordinate basis $\{\mathbf{e}_i\}$. In any motion of $\mathscr{Q}$ the quantities $(x^{\alpha,i}, \xi^{\beta,i}, \theta^{\beta,i})$ are all functions of time. Before imposing any constraints on $\mathscr{Q}$, we can regard a configuration of $\mathscr{Q}$ simply as a point in $\mathscr{R}^{3N} \times \mathscr{R}^{3K} \times \mathscr{SO}(3)^K$, which is a manifold of dimension $3N + 6K$ in the Euclidean space $\mathscr{R}^{3N} \times \mathscr{R}^{3K} \times \mathscr{R}^{9K}$. Hence we call the manifold $\mathscr{R}^{3N} \times \mathscr{R}^{3K} \times \mathscr{SO}(3)^K$ the *free configuration space* of $\mathscr{Q}$, and we regard $(x^{\alpha,i}, \xi^{\beta,i}, \theta^{\beta,i})$ as the *standard coordinates* on the free configuration space.

We define a *time-independent holonomic constraint* on $\mathscr{Q}$ by a smooth surface $\mathscr{M}$ in the free configuration space, and we require that the configuration of $\mathscr{Q}$ be constrained to stay in $\mathscr{M}$. Hence we call $\mathscr{M}$ the *constraint configuration space* of $\mathscr{Q}$. The dimension n of $\mathscr{M}$ is then called the *number of degrees of freedom* of the constrained system. As explained in Section 68, IVT-2, we can characterize the constraint surface $\mathscr{M}$ by a system of algebraic equations

$$\begin{aligned} x^{\alpha,i} &= \varphi^{\alpha,i}(q^1, \ldots, q^n), \\ \xi^{\beta,i} &= \psi^{\beta,i}(q^1, \ldots, q^n), \\ \theta^{\beta,i} &= \zeta^{\beta,i}(q^1, \ldots, q^n), \end{aligned} \tag{3.2}$$

where $\varphi^{\alpha,i}, \psi^{\beta,i}, \zeta^{\beta,i}$ are certain smooth functions, and where $q^1, \ldots, q^n$ are certain surface coordinates in $\mathscr{M}$. Using the representation (3.2), we can identify a point in the constraint space $\mathscr{M}$ by (q^Δ). Hence we call $(q^\Delta, \Delta = 1, \ldots, n)$ the *generalized coordinates* for the constrained system $\mathscr{Q}$. Since the configuration of $\mathscr{Q}$ must stay in $\mathscr{M}$, a motion of $\mathscr{Q}$ may be described by a system of equations of the form

$$q^\Delta = q^\Delta(t), \qquad \Delta = 1, \ldots, n. \tag{3.3}$$

By using the convention as explained in the preceding two sections, we suppress the argument t and denote the motion of $\mathscr{Q}$ simply by (q^Δ). We shall now develop the kinematics of the constrained system $\mathscr{Q}$ in detail.

We remark first that since the dimension of $\mathscr{M}$ is n, the generalized coordinates (q^Δ) must be independent, i.e., the tangent vectors of the coordinate curves of (q^Δ) must be linearly independent and span a vector space of dimension n at any point of $\mathscr{M}$. Specifically, let $\mathbf{q}$ be a point in $\mathscr{M}$. We denote the tangent vector of the q^Δ coordinate curve at $\mathbf{q}$ by $\mathbf{h}_\Delta(\mathbf{q})$. Then we define

$$\mathscr{M}_{\mathbf{q}} \equiv \operatorname{span}\{\mathbf{h}_\Delta(\mathbf{q}), \Delta = 1, \ldots, n\}. \tag{3.4}$$

Since $\{\mathbf{h}_\Delta\}$ is a linearly independent set, $\mathscr{M}_\mathbf{q}$ is an n-dimensional vector space, called the *tangent space* of $\mathscr{M}$ at $\mathbf{q}$.

The tangent space $\mathscr{M}_\mathbf{q}$, which we have just defined, is a very important concept, so we now explain the details of the steps leading to its definition. First, the free configuration space $\mathscr{R}^{3N}\times\mathscr{R}^{3K}\times\mathscr{SO}(3)^K$ is a smooth surface in the Euclidean space $\mathscr{R}^{3N}\times\mathscr{R}^{3K}\times\mathscr{R}^{9K}$. As usual, we use the standard Cartesian coordinate system $(x^{\alpha,i}, \xi^{\beta,i}, Q^\beta_{ij})$. Then the equation (2.25) defines the free configuration space $\mathscr{R}^{3N}\times\mathscr{R}^{3K}\times\mathscr{SO}(3)^K \subset \mathscr{R}^{3N}\times\mathscr{R}^{3K}\times\mathscr{R}^{9K}$ in terms of the surface coordinates $(x^{\alpha,i}, \xi^{\beta,i}, \theta^{\beta,i})$. The tangent vectors of the coordinate curves of $x^{\alpha,i}$, $\xi^{\beta,i}$, and $\theta^{\beta,i}$ are linearly independent in $\mathscr{R}^{3N}\times\mathscr{R}^{3K}\times\mathscr{R}^{9K}$ and span a tangent space of dimension $3N+6K$ at each point of $\mathscr{R}^{3N}\times\mathscr{R}^{3K}\times\mathscr{SO}(3)^K$.

Similarly, the system (3.2) defines the constraint space $\mathscr{M}\subset\mathscr{R}^{3N}\times\mathscr{R}^{3K}\times\mathscr{SO}(3)^K$. The tangent vectors of the coordinate curves of (q^Δ) are linearly independent in the tangent space of $\mathscr{R}^{3N}\times\mathscr{R}^{3K}\times\mathscr{SO}(3)^K$ and span a tangent space $\mathscr{M}_\mathbf{q}$ at each point $\mathbf{q}$ in $\mathscr{M}$. Of course, we can regard $\mathscr{M}$ as a surface in the Euclidean space $\mathscr{R}^{3N}\times\mathscr{R}^{3K}\times\mathscr{R}^{9K}$ also. Then $\mathscr{M}_\mathbf{q}$ is an n-dimensional subspace of $\mathscr{R}^{3N}\times\mathscr{R}^{3K}\times\mathscr{R}^{9K}$. It should be noted, however, that the subspace $\mathscr{M}_\mathbf{q}$ depends on the point $\mathbf{q}\in\mathscr{M}$. Indeed, if $\mathbf{q}$ and $\bar{\mathbf{q}}$ are two different points in $\mathscr{M}$, then generally $\mathscr{M}_\mathbf{q}\neq\mathscr{M}_{\bar{\mathbf{q}}}$.

If we denote the tangent vectors of the coordinate curves of $x^{\alpha,i}$, $\xi^{\beta,i}$, and $\theta^{\beta,i}$ by $\mathbf{u}_{\alpha,i}$, $\mathbf{v}_{\beta,i}$, and $\mathbf{w}_{\beta,i}$, respectively, then the system (3.2) gives the following component formula for the tangent vector $\mathbf{h}_\Delta$ in terms of the basis $\{\mathbf{u}_{\alpha,i}, \mathbf{v}_{\beta,i}, \mathbf{w}_{\beta,i}\}$:

$$\mathbf{h}_\Delta = \frac{\partial\varphi^{\alpha,i}}{\partial q^\Delta}\mathbf{u}_{\alpha,i} + \frac{\partial\psi^{\beta,i}}{\partial q^\Delta}\mathbf{v}_{\beta,i} + \frac{\partial\zeta^{\beta,i}}{\partial q^\Delta}\mathbf{w}_{\beta,i}, \tag{3.5}$$

where the repeated indices (α, i) and (β, i) are summed over their appropriate ranges: α, 1 to N, β, 1 to K, and i, 1 to 3. The condition that $\{\mathbf{h}_\Delta\}$ is linearly independent corresponds to the requirement that the $(3N+6K)\times n$ coefficient matrix

$$\left[\frac{\partial\varphi^{\alpha,i}}{\partial q^\Delta}, \frac{\partial\psi^{\beta,i}}{\partial q^\Delta}, \frac{\partial\zeta^{\beta,i}}{\partial q^\Delta}\right]$$

in (3.5) has rank n at each point $\mathbf{q}\in\mathscr{M}$. This condition is the same as the condition of independence in multivariable calculus for the parametric representations of $x^{\alpha,i}$, $\xi^{\beta,i}$, and $\theta^{\beta,i}$ by (q^Δ) as shown in (3.2).

Now consider a motion of $\mathscr{Q}$ given by (3.3). As explained in Section 1, we can regard such a motion as a curve in $\mathscr{M}$. The tangent vector of the curve is a vector $\dot{\mathbf{q}} \in \mathscr{M}_{\mathbf{q}}$ with component form

$$\dot{\mathbf{q}} = \frac{dq^{\Delta}}{dt} \mathbf{h}_{\Delta}(\mathbf{q}) \tag{3.6}$$

relative to the natural basis $\{\mathbf{h}_{\Delta}\}$ at $\mathbf{q}$. In (3.6) the repeated index Δ is summed from 1 to n, of course. We call the tangent vector $\dot{\mathbf{q}}$ the *generalized velocity* of the motion. This vector is the basic kinematic quantity for the system $\mathscr{Q}$.

From the vector $\dot{\mathbf{q}}$ we can derive various other kinematical quantities for the particles p^{α} and for the rigid bodies $\mathscr{B}^{\beta}$ in the system $\mathscr{Q}$. Specifically, from (3.2a) the ordinary velocity $\mathbf{v}^{\alpha} = v^{\alpha,i}\mathbf{e}_i$ of p^{α} is given by

$$v^{\alpha,i} = \frac{\partial \varphi^{\alpha,i}}{\partial q^{\Delta}} \frac{dq^{\Delta}}{dt}, \tag{3.7}$$

which means that $\mathbf{v}^{\alpha}$ is the image of $\dot{\mathbf{q}}$ under a linear map grad $\boldsymbol{\varphi}^{\alpha}: \mathscr{M}_{\mathbf{q}} \to \mathscr{V}$, viz.,

$$\mathbf{v}^{\alpha} = [\text{grad}\, \boldsymbol{\varphi}^{\alpha}](\dot{\mathbf{q}}). \tag{3.8}$$

The component form of this linear map relative to the bases $\{\mathbf{e}_i\}$ in $\mathscr{V}$ and $\{\mathbf{h}_{\Delta}(\mathbf{q})\}$ in $\mathscr{M}_{\mathbf{q}}$ is given by (3.7). Similarly, the ordinary velocity $\mathbf{V}^{\beta} = V^{\beta,i}\mathbf{e}_i$ of the center of mass $p^{\beta,c}$ of the rigid body $\mathscr{B}^{\beta}$ is given by

$$V^{\beta,i} = \frac{\partial \psi^{\beta,i}}{\partial q^{\Delta}} \frac{dq^{\Delta}}{dt}, \tag{3.9}$$

which again means that $\mathbf{V}^{\beta}$ is the image of $\dot{\mathbf{q}}$ under a linear map grad $\boldsymbol{\psi}^{\beta}$: $\mathscr{M}_{\mathbf{q}} \to \mathscr{V}$, viz.,

$$\mathbf{V}^{\beta} = [\text{grad}\, \boldsymbol{\psi}^{\beta}](\dot{\mathbf{q}}). \tag{3.10}$$

Finally, the rate of change of the Eulerian angle $\theta^{\beta,i}$ of the imbedded basis $\{\mathbf{f}_i^{\beta}\}$ of $\mathscr{B}^{\beta}$ relative to the fixed basis $\{\mathbf{e}_i\}$ is given by

$$\dot{\theta}^{\beta,i} = \frac{\partial \zeta^{\beta,i}}{\partial q^{\Delta}} \frac{dq^{\Delta}}{dt}. \tag{3.11}$$

Now by using the transformation (2.28), we obtain the angular velocity $\boldsymbol{\omega}^{\beta} = \omega^{\beta,i}\mathbf{f}_i^{\beta}$ of $\mathscr{B}^{\beta}$ by

$$\omega^{\beta,i} = W^{\beta,ij}\dot{\theta}^{\beta,j} = W^{\beta,ij} \frac{\partial \zeta^{\beta,j}}{\partial q^{\Delta}} \frac{dq^{\Delta}}{dt}, \qquad (\beta \text{ not summed}) \tag{3.12}$$

which means that the angular velocity $\boldsymbol{\omega}^{\beta}$ is the image of $\dot{\mathbf{q}}$ under a linear map $\mathbf{W}^\beta \circ \operatorname{grad} \boldsymbol{\zeta}^\beta: \mathscr{M}_\mathbf{q} \to \mathscr{V}$, viz.,

$$\boldsymbol{\omega}^\beta = [\mathbf{W}^\beta \circ \operatorname{grad} \boldsymbol{\zeta}^\beta](\dot{\mathbf{q}}). \tag{3.13}$$

Having obtained the velocities $\mathbf{v}^\alpha$, $\mathbf{V}^\beta$, and the angular velocities $\boldsymbol{\omega}^\beta$ in terms of $\dot{\mathbf{q}}$ for all particles and rigid bodies in the system $\mathscr{Q}$, we can use the results of the preceding two sections to determine other kinematical quantities. As we shall see, the total kinetic energy E of $\mathscr{Q}$ turns out to be a very important kinematical quantity in the derivation of Lagrange's equations. Therefore we now derive a representation for E in terms of $\dot{\mathbf{q}}$.

First, from (1.13c) and (3.7) the kinetic energy e^α of p^α is given by

$$e^\alpha = \frac{1}{2} m^\alpha \frac{\partial \varphi^{\alpha,i}}{\partial q^\Delta} \frac{\partial \varphi^{\alpha,i}}{\partial q^\Gamma} \frac{dq^\Delta}{dt} \frac{dq^\Gamma}{dt}, \qquad \alpha = 1, \ldots, N, \tag{3.14}$$

where m^α denotes the mass of p^α. As we have indicated in (3.14), α is a free index. The representation (3.14) shows clearly that e^α is given by a quadratic function of $\dot{\mathbf{q}}$. Next, from (2.40), (3.10), and (3.12) the kinetic energy E^β of $\mathscr{B}^\beta$ is given by

$$E^\beta = \frac{1}{2}\left[M^\beta \frac{\partial \psi^{\beta,i}}{\partial q^\Delta} \frac{\partial \psi^{\beta,i}}{\partial q^\Gamma} + W^{\beta,ij} \frac{\partial \zeta^{\beta,j}}{\partial q^\Delta} I^{\beta,ik} W^{\beta,kl} \frac{\partial \zeta^{\beta,l}}{\partial q^\Gamma}\right] \frac{dq^\Delta}{dt} \frac{dq^\Gamma}{dt}, \tag{3.15}$$

where β is a free index, i.e., $\beta = 1, \ldots, K$, and where M^β and $I^{\beta,ij}$ are the mass and the inertia tensor relative to the center of mass of $\mathscr{B}^\beta$. The total energy E is then given by the representation

$$\begin{aligned} E = \frac{1}{2}\Bigg[& m^\alpha \frac{\partial \varphi^{\alpha,i}}{\partial q^\Delta} \frac{\partial \varphi^{\alpha,i}}{\partial q^\Gamma} + M^\beta \frac{\partial \psi^{\beta,i}}{\partial q^\Delta} \frac{\partial \psi^{\beta,i}}{\partial q^\Gamma} \\ & + W^{\beta,ij} \frac{\partial \zeta^{\beta,j}}{\partial q^\Delta} I^{\beta,ik} W^{\beta,kl} \frac{\partial \zeta^{\beta,l}}{\partial q^\Gamma}\Bigg] \frac{dq^\Delta}{dt} \frac{dq^\Gamma}{dt} \\ \equiv {} & \frac{1}{2} g_{\Delta\Gamma} \frac{dq^\Delta}{dt} \frac{dq^\Gamma}{dt}, \end{aligned} \tag{3.16}$$

where all the repeated indices are summed over their appropriate ranges: α, 1 to N; β, 1 to K; i, j, k, l, 1 to 3; Δ and Γ, 1 to n. The formula (3.16) shows that in general E is given by a quadratic function of $\dot{\mathbf{q}}$.

The coefficient matrix $[g_{\Delta\Gamma}]$ on the right-hand side of (3.16) is a function of position $\mathbf{q} \in \mathscr{M}$ and possesses some very important properties:

(i) $[g_{\Delta\Gamma}]$ is symmetric; i.e., $g_{\Delta\Gamma} = g_{\Gamma\Delta}$, $\Delta, \Gamma = 1, \ldots, n$. This property follows directly from (3.16) and the symmetry condition of the inertia tensor.

(ii) $[g_{\Delta\Gamma}]$ is positive definite; i.e., $g_{\Delta\Gamma}\eta^\Delta\eta^\Gamma > 0$ unless $\eta^\Delta = 0$ for all $\Delta = 1, \ldots, n$. This property is a consequence of the facts that m^α and M^β are positive, that $[I^{\beta,ij}]$ are positive definite (or, equivalently, that $[W^{\beta,ij}I^{\beta,ik}W^{\beta,kl}]$ are positive definite, since $[W^{\beta,ij}]$ are nonsingular), and that $\{\mathbf{h}_\Delta\}$ is linearly independent. Indeed, from (3.16) the quadratic form $g_{\Delta\Gamma}\eta^\Delta\eta^\Gamma$ can be rewritten as

$$g_{\Delta\Gamma}\eta^\Delta\eta^\Gamma = m^\alpha\lambda^{\alpha,i}\lambda^{\alpha,i} + M^\beta\varkappa^{\beta,i}\varkappa^{\beta,i} + W^{\beta,ij}I^{\beta,ik}W^{\beta,kl}\tau^{\beta,j}\tau^{\beta,l}, \tag{3.17}$$

where

$$\lambda^{\alpha,i} = \frac{\partial\varphi^{\alpha,i}}{\partial q^\Delta}\eta^\Delta, \qquad \varkappa^{\beta,i} = \frac{\partial\psi^{\beta,i}}{\partial q^\Delta}\eta^\Delta, \qquad \tau^{\beta,i} = \frac{\partial\zeta^{\beta,i}}{\partial q^\Delta}\eta^\Delta. \tag{3.18}$$

By using the facts that m^α and M^β are positive, and that $[W^{\beta,ij}I^{\beta,ik}W^{\beta,kl}]$ are positive definite, we obtain from (3.17) the result $g_{\Delta\Gamma}\eta^\Delta\eta^\Gamma > 0$, unless $(\lambda^{\alpha,i}, \varkappa^{\beta,i}, \tau^{\beta,i})$ all vanish. Now by virtue of the fact that $\{\mathbf{h}_\Delta\}$ is linearly independent, the matrix

$$\left[\frac{\partial\varphi^{\alpha,i}}{\partial q^\Delta}, \frac{\partial\psi^{\beta,i}}{\partial q^\Delta}, \frac{\partial\zeta^{\beta,i}}{\partial q^\Delta}\right]$$

has rank n. Consequently, from (3.18) $(\lambda^{\alpha,i}, \varkappa^{\beta,i}, \tau^{\beta,i})$ all vanish if and only if (η^Δ) all vanish. Thus the proof of (ii) is complete.

(iii) Under a change of generalized coordinates the matrix $[g_{\Delta\Gamma}]$ transforms according to the transformation rule of the component matrix of a second-order covariant tensor. To prove this property we consider a change of coordinates from (q^Δ) to $(\bar{q}^\Delta)$ on $\mathscr{M}$. The coordinate transformation is given by

$$q^\Delta = q^\Delta(\bar{q}^1, \ldots, \bar{q}^n), \qquad \bar{q}^\Delta = \bar{q}^\Delta(q^1, \ldots, q^n), \qquad \Delta = 1, \ldots, n. \tag{3.19}$$

Substituting (3.19) into (3.2), we obtain the system

$$\begin{aligned} x^{\alpha,i} &= \bar{\varphi}^{\alpha,i}(\bar{q}^1, \ldots, \bar{q}^n), \\ \xi^{\beta,i} &= \bar{\psi}^{\beta,i}(\bar{q}^1, \ldots, \bar{q}^n), \\ \theta^{\beta,i} &= \bar{\zeta}^{\beta,i}(\bar{q}^1, \ldots, \bar{q}^n), \end{aligned} \tag{3.20}$$

which characterizes $\mathscr{M}$ in terms of $(\bar{q}^\Delta)$. The matrix $[\bar{g}_{\Delta\Gamma}]$ relative to $(\bar{q}^\Delta)$ can be obtained from (3.16) by replacing $\varphi^{\alpha,i}$, $\psi^{\beta,i}$, $\zeta^{\beta,i}$, and q^Δ by $\bar{\varphi}^{\alpha,i}$, $\bar{\psi}^{\beta,i}$, $\bar{\zeta}^{\beta,i}$, and $\bar{q}^\Delta$, respectively. But by the chain rule based on (3.19) the partial derivatives of the functions $\varphi^{\alpha,i}$, $\bar{\varphi}^{\alpha,i}$, $\psi^{\beta,i}$, $\bar{\psi}^{\beta,i}$, $\zeta^{\beta,i}$, $\bar{\zeta}^{\beta,i}$ are related by

$$\frac{\partial\bar{\varphi}^{\alpha,i}}{\partial\bar{q}^\Delta} = \frac{\partial\varphi^{\alpha,i}}{\partial q^\Gamma}\frac{\partial q^\Gamma}{\partial\bar{q}^\Delta}, \qquad \frac{\partial\bar{\psi}^{\beta,i}}{\partial\bar{q}^\Delta} = \frac{\partial\psi^{\beta,i}}{\partial q^\Gamma}\frac{\partial q^\Gamma}{\partial\bar{q}^\Delta}, \qquad \frac{\partial\bar{\zeta}^{\beta,i}}{\partial\bar{q}^\Delta} = \frac{\partial\zeta^{\beta,i}}{\partial q^\Gamma}\frac{\partial q^\Gamma}{\partial\bar{q}^\Delta}. \tag{3.21}$$

Substituting these relations into the definition for $\bar{g}_{\Delta\Gamma}$ and $g_{\Delta\Gamma}$, we obtain immediately

$$\bar{g}_{\Delta\Gamma} = g_{\Lambda\Xi} \frac{\partial q^{\Lambda}}{\partial \bar{q}^{\Delta}} \frac{\partial q^{\Xi}}{\partial \bar{q}^{\Gamma}}. \tag{3.22}$$

Thus the proof of (iii) is complete.

The preceding three properties are equivalent to the assertion that $[g_{\Delta\Gamma}(\mathbf{q})]$ is the component matrix of an inner product $\mathbf{g}(\mathbf{q})$ on the tangent space $\mathscr{M}_{\mathbf{q}}$. Since $[g_{\Delta\Gamma}(\mathbf{q})]$ depends smoothly on q^{Δ}, the field $\mathbf{g}$ is a Riemannian metric on the manifold $\mathscr{M}$. We call $\mathbf{g}$ the *inertia metric*. From (3.16) the total energy E of $\mathscr{Q}$ is just $\frac{1}{2} \| \dot{\mathbf{q}} \|_{\mathbf{g}}^2$, where the norm is taken relative to the inertia metric $\mathbf{g}$, viz.,

$$\| \dot{\mathbf{q}} \|_{\mathbf{g}}^2 = \mathbf{g}(\dot{\mathbf{q}}, \dot{\mathbf{q}}). \tag{3.23}$$

As explained in Section 56, IVT-2, the metric $\mathbf{g}$ gives rise to an operation of covariant derivative, which enables us to take the covariant (time) derivative of a vector field along any curve in $\mathscr{M}$. In particular, it becomes meaningful to take the rate of change of the generalized velocity $\dot{\mathbf{q}}$ along the curve $\mathbf{q} = \mathbf{q}(t)$. This time derivative corresponds to the *generalized acceleration*, which we have not yet defined in the context of the constrained system $\mathscr{Q}$.

So far, we have developed the kinematics of a constrained system $\mathscr{Q}$ subject to a time-independent holonomic constraint. A more general type of constraint is a *time-dependent holonomic constraint*, which is defined by a moving surface $\mathscr{M}^t$ in the free configuration space. To characterize $\mathscr{M}^t$, we use a system of algebraic equations of the form

$$\begin{aligned} x^{\alpha,i} &= \varphi^{\alpha,i}(q^1, \ldots, q^n, t), \\ \xi^{\beta,i} &= \psi^{\beta,i}(q^1, \ldots, q^n, t), \\ \theta^{\beta,i} &= \zeta^{\beta,i}(q^1, \ldots, q^n, t), \end{aligned} \tag{3.24}$$

where the functions $\varphi^{\alpha,i}$, $\psi^{\beta,i}$, $\zeta^{\beta,i}$ may depend explicitly on t. We can represent a motion of $\mathscr{Q}$ still by the system (3.3). However, the curve $\mathbf{q} = \mathbf{q}(t)$ now lies in the manifold

$$\tilde{\mathscr{M}} = \bigcup_{t \in \mathscr{R}} \mathscr{M}^t, \tag{3.25}$$

which generally has dimension $n + 1$.

We define the tangent vector $\mathbf{h}_\Delta$ of the q^Δ coordinate curve again by (3.5), but we need also the vector

$$\mathbf{h}_{n+1} \equiv \frac{\partial \varphi^{\alpha,i}}{\partial t} \mathbf{u}_{\alpha,i} + \frac{\partial \psi^{\beta,i}}{\partial t} \mathbf{v}_{\beta,i} + \frac{\partial \zeta^{\beta,i}}{\partial t} \mathbf{w}_{\beta,i}. \tag{3.26}$$

The tangent vector of the curve $\mathbf{q} = \mathbf{q}(t)$ now has the component form

$$\dot{\mathbf{q}} = \frac{dq^\Delta}{dt} \mathbf{h}_\Delta + \mathbf{h}_{n+1}. \tag{3.27}$$

We call $\dot{\mathbf{q}}$ the *generalized velocity* again. From $\dot{\mathbf{q}}$ we can calculate the velocity $\mathbf{v}^\alpha = v^{\alpha,i}\mathbf{e}_i$ of p^α by

$$v^{\alpha,i} = \frac{\partial \varphi^{\alpha,i}}{\partial q^\Delta} \frac{dq^\Delta}{dt} + \frac{\partial \varphi^{\alpha,i}}{\partial t}, \tag{3.28}$$

the velocity $\mathbf{V}^\beta = V^{\beta,i}\mathbf{e}_i$ of the center of mass of $\mathscr{B}^\beta$ by

$$V^{\beta,i} = \frac{\partial \psi^{\beta,i}}{\partial q^\Delta} \frac{dq^\Delta}{dt} + \frac{\partial \psi^{\beta,i}}{\partial t}, \tag{3.29}$$

the rate of change of the Eulerian angle $\theta^{\beta,i}$ of $\mathscr{B}^\beta$ by

$$\dot{\theta}^{\beta,i} = \frac{\partial \zeta^{\beta,i}}{\partial q^\Delta} \frac{dq^\Delta}{dt} + \frac{\partial \zeta^{\beta,i}}{\partial t}, \tag{3.30}$$

and, finally, the angular velocity $\boldsymbol{\omega}^\beta = \omega^{\beta,i}\mathbf{f}_i^\beta$ of $\mathscr{B}^\beta$ by

$$\omega^{\beta,i} = W^{\beta,ij}\dot{\theta}^{\beta,j} = W^{\beta,ij}\left(\frac{\partial \zeta^{\beta,j}}{\partial q^\Delta} \frac{dq^\Delta}{dt} + \frac{\partial \zeta^{\beta,j}}{\partial t} \right). \tag{3.31}$$

In the regular case when $\mathscr{M}$ has dimension $n + 1$, the set $\{\mathbf{h}_1, \ldots, \mathbf{h}_n, \mathbf{h}_{n+1}\}$ is linearly independent and forms a basis for the tangent space $\mathscr{M}_\mathbf{q}$. Then $\dot{\mathbf{q}} \in \mathscr{M}_\mathbf{q}$, and we can regard (3.28), (3.29), and (3.31) as the component forms of certain linear maps from $\mathscr{M}_\mathbf{q}$ to $\mathscr{V}$ sending $\dot{\mathbf{q}}$ to the vectors $\mathbf{v}^\alpha$, $\mathbf{V}^\beta$, and $\boldsymbol{\omega}^\beta$, respectively. We can regard t as the generalized coordinate q^{n+1} on $\mathscr{M}$. Then a motion of $\mathscr{Q}$ is given by a curve in $\mathscr{M}$ having the special coordinate form

$$q^\Delta = q^\Delta(t), \qquad q^{n+1} = t. \tag{3.32}$$

In this sense the formulas (3.27), (3.28), (3.29), (3.30), and (3.31) are the same as (3.6), (3.7), (3.9), (3.11), and (3.12), respectively, when the summation on Δ is extended to $(n + 1)$. In particular, the total energy E is still given by (3.16) except that Δ, Γ are summed from 1 to $(n + 1)$.

4. Dynamical Principles for Particles and Rigid Bodies

In the preceding three sections we have considered motions in general for particles and rigid bodies. In classical mechanics motions of particles and rigid bodies are regarded as results caused by forces and moments and are governed by certain principles of dynamics. We develop first the dynamical principles for a particle p.

When p undergoes a motion $\mathbf{x} = \mathbf{x}(t) \in \mathscr{S}$, a force system acting on p is just a vector $\mathbf{f} = \mathbf{f}(t) \in \mathscr{V}$ assigned at the position $\mathbf{x}(t)$ for each $t \in \mathscr{R}$. Unlike a kinematical quantity, a force is actually a vector defined directly in the instantaneous space and is, therefore, independent of the choice of any frame of reference. If a particular frame of reference is used, we can represent the force by a vector in the translation space of the frame. However, since the representation of the instantaneous space by the physical space of the frame is required to be an isometry, the magnitude of the force remains invariant under any change of frames. Therefore if the force vanishes relative to any one frame, then it vanishes relative to all frames. Clearly, the velocity or the acceleration do not enjoy such an invariance property. Indeed, given any motion $\{\mathbf{x}_\tau, \tau \in \mathscr{T}\}$ in the Newtonian space–time $\mathscr{E}$, we can always reduce the velocity and the acceleration to zero by choosing the origin of the frame at $\mathbf{x}_\tau$, $\tau \in \mathscr{T}$. Relative to some other frames, the velocity and the acceleration of this motion need not vanish of course.

We now state the dynamical principles for a particle.

Newton's First Law. There exists a particular frame of reference, called an *inertial frame*, relative to which the linear momentum of a particle remains constant when the force acting on the particle vanishes.

As remarked before, the condition that force vanishes is independent of the choice of the frame. Consequently, the condition that linear momentum remains constant under vanishing force can be used as a criterion for an inertial frame, and Newton's first law merely asserts the existence of such a frame. An inertial frame is not unique, of course. We shall now choose a particular inertial frame as the frame of reference.

Newton's Second Law. Relative to an inertial frame the motion of a particle satisfies the following *equation of motion*:

$$\frac{d}{dt}\mathbf{l}(t) = \mathbf{f}(t), \qquad t \in \mathscr{R}. \tag{4.1}$$

Clearly this principle is consistent with the first principle, since from (4.1) $\mathbf{f} = \mathbf{0}$ implies $\mathbf{l} =$ const. This remark does not mean that the first principle is a consequence of the second principle, however, since without the existence of an inertial frame, the equation of motion is meaningless. In fact the two sides of (4.1) cannot possibly be equal in all frames, since the right-hand side is frame indifferent, while the left-hand side is not.

From (1.14) the equation of motion (4.1) can be rewritten as

$$m\mathbf{a}(t) = m\,\frac{d^2\mathbf{x}(t)}{dt^2} = \mathbf{f}(t), \qquad t \in \mathscr{R}. \tag{4.2}$$

Relative to the fixed rectangular Cartesian coordinate system (x^i) in $\mathscr{S}$, (4.2) corresponds to a system of second-order differential equations

$$m\,\frac{d^2x^i(t)}{dt^2} = f^i(t), \qquad t \in \mathscr{R}, \qquad i = 1, 2, 3, \tag{4.3}$$

for the coordinate functions $x^i(t)$ of the motion. We can solve the system (4.3) and determine $x^i(t)$ when $f^i(t)$ and certain initial values of $x^i(t)$ are known. In applications $f^i(t)$ are not always known, however. For example, the particle p may be an element in a certain constrained system. Then the force $\mathbf{f}$ is the resultant of a certain known external force and a certain unknown constraint force. The value of the latter may be determined by the condition that the coordinate functions $x^i(t)$ satisfy the constraint. For such problems it is more convenient to characterize the motion by the generalized coordinates. We shall derive the equations of motion in terms of the generalized coordinates in Section 5. These equations are *Lagrange's equations*.

Having developed the dynamical principles for a particle, we consider next the same for a rigid body. Unlike a particle, a rigid body in general may be acted on by a force system consisting in a resultant force $\mathbf{F} = \mathbf{F}(t) \in \mathscr{V}$ and a resultant moment $\mathbf{G} = \mathbf{G}(t) \in \mathscr{V}$ relative to the center of mass at each $t \in \mathscr{R}$. As before both $\mathbf{F}$ and $\mathbf{G}$ are independent of the frame of reference. Hence in order that $\mathbf{F}$ and $\mathbf{G}$ may be related to certain kinematical quantities, a particular frame must be used.

We now state the dynamical principles for a rigid body $\mathscr{B}$.

Euler's First Law. Relative to an inertial frame the rate of change of the linear momentum of $\mathscr{B}$ is equal to the resultant force acting on $\mathscr{B}$.

We can express this principle by the *linear momentum equation*:

$$\frac{d}{dt}\mathbf{L}(t) = \mathbf{F}(t), \qquad t \in \mathscr{R}. \tag{4.4}$$

Clearly, this equation is consistent with the equation of motion (4.1). Using (2.30), we can rewrite (4.4) as

$$M\mathbf{a}^c(t) = M\frac{d^2\mathbf{x}^c(t)}{dt^2} = \mathbf{F}(t), \qquad t \in \mathscr{R}, \tag{4.5}$$

which has exactly the same form as (4.2). Thus the motion of the center of mass of a rigid body is just like that of a particle with mass M and acted on by the force $\mathbf{F}$.

Euler's Second Law (fixed-center-of-moment version). Relative to the origin of an inertial frame the rate of change of the moment of momentum of $\mathscr{B}$ is equal to the total moment $\mathbf{r}^c \times \mathbf{F} + \mathbf{G}$ acting on $\mathscr{B}$.

We can express this principle by the *moment of momentum equation* (relative to the fixed origin)

$$\frac{d\mathbf{H}}{dt} = \mathbf{r}^c \times \mathbf{F} + \mathbf{G}, \tag{4.6}$$

where the first term on the right-hand side denotes the moment of the resultant force relative to the origin, and the second term denotes the moment (a couple) relative to the center of mass. From (2.36) and (4.5) we can rewrite (4.6) in the form

$$\frac{d}{dt}(\mathbf{I}\boldsymbol{\omega}) = \mathbf{G}. \tag{4.7}$$

This equation is also called the *moment of momentum equation* (relative to the center of mass), and it can be stated as Euler's second law.

Euler's Second Law (center-of-mass version). Relative to the center of mass the rate of change of the moment of momentum of $\mathscr{B}$ in an inertial frame is equal to the total moment $\mathbf{G}$ acting on $\mathscr{B}$.

Substituting (2.37) into (4.7), we obtain

$$\mathbf{I}\dot{\boldsymbol{\omega}} + \boldsymbol{\omega} \times \mathbf{I}\boldsymbol{\omega} = \mathbf{G}. \tag{4.8}$$

In component form relative to an imbedded basis $\{\mathbf{f}_i\}$ at the center of mass this equation is given by

$$I^{ij}\frac{d\omega^j}{dt}+\varepsilon^{ijk}\omega^j I^{kl}\omega^l=G^i,\qquad i=1,2,3. \tag{4.9}$$

In particular, if $\{\mathbf{f}_i\}$ is taken to be a principal basis of the inertia tensor $\mathbf{I}$, then (4.9) reduces to

$$\begin{aligned} I_1\frac{d\omega^1}{dt}-(I_2-I_3)\omega^2\omega^3&=G^1,\\ I_2\frac{d\omega^2}{dt}-(I_3-I_1)\omega^3\omega^1&=G^2,\\ I_3\frac{d\omega^3}{dt}-(I_1-I_2)\omega^1\omega^2&=G^3, \end{aligned} \tag{4.10}$$

where I_1, I_2, I_3 are the (positive) proper numbers of $\mathbf{I}$. The equations (4.10) are *Euler's equations*.

Relative to a fixed rectangular Cartesian coordinate system (x^i) in $\mathscr{S}$ the linear momentum equation (4.5) corresponds to a system of second-order differential equations

$$M\frac{d^2\xi^i(t)}{dt^2}=F^i(t),\qquad t\in\mathscr{R},\qquad i=1,2,3, \tag{4.11}$$

where the components of $\mathbf{F}$ are taken relative to the fixed coordinate basis $\{\mathbf{e}_i\}$. These equations govern the first three parameters, ξ^1, ξ^2, ξ^3 of the motion of $\mathscr{B}$ as functions of t.

Relative to the same fixed coordinate system (x^i) the moment of momentum equation (4.8) corresponds to a system of second-order differential equations

$$I^{ij}W^{jk}\frac{d^2\theta^k}{dt^2}+\left(I^{ij}\frac{\partial W^{jk}}{\partial\theta^l}+\varepsilon^{ijr}W^{jl}I^{rs}W^{sk}\right)\frac{d\theta^l}{dt}\frac{d\theta^k}{dt}=G^i, \tag{4.12}$$

where I^{ij}, G^i, and W^{jk} are taken relative to the imbedded basis $\{\mathbf{f}_i\}$ at the center of mass. These equations govern the last three parameters, θ^1, θ^2, θ^3 of the motion of $\mathscr{B}$ as functions of t.

Like the system (4.3), the systems (4.11) and (4.12) can be used to determine the motion provided that $F^i(t)$ and $G^i(t)$, and certain initial values of $\xi^i(t)$ and $\theta^i(t)$, are known. Such is not the case, however, when $\mathscr{B}$ is an element of a constrained system. The reason has been explained before, since the constraint force and moment acting on $\mathscr{B}$ are generally not known *a priori*, and their values are determined by the condition that the parameters $\xi^i(t)$ and $\theta^i(t)$ satisfy the constraint. As we shall see in the following section, under certain assumptions this difficulty can be overcome by using the generalized coordinates and their governing Lagrange's equations.

5. Lagrange's Equations for Constrained Systems

The concept of a constrained system $\mathscr{Q}$ consisting in particles $\{p^\alpha\}$ and rigid bodies $\{\mathscr{B}^\beta\}$ was introduced in Section 3. We have shown that a motion of $\mathscr{Q}$ can be characterized by the system (3.3), $q^\Delta = q^\Delta(t)$, and we have defined various kinematical quantities in terms of $q^\Delta(t)$. When the generalized coordinates $q^\Delta(t)$ are determined, the standard coordinates $x^{\alpha,i}(t)$, $\xi^{\beta,i}(t)$, $\theta^{\beta,i}(t)$ are given by the constraint equations (3.2) or (3.24). In this section we shall derive the dynamical governing equations for the generalized coordinates $q^\Delta(t)$. These equations are known as *Lagrange's equations*.

The starting point of the derivation for Lagrange's equations is the equations of linear momentum (4.3) for p^α and (4.11) for $\mathscr{B}^\beta$ and the equations of moment of momentum (4.12) for $\mathscr{B}^\beta$. These equations govern the dependence of the standard coordinates $x^{\alpha,i}$, $\xi^{\beta,i}$, $\theta^{\beta,i}$ on time in the free configuration space $\mathscr{R}^{3N}\times\mathscr{R}^{3K}\times\mathscr{SO}(3)^K$. As remarked in the preceding section, for a constrained system the forces acting on the elements of the system generally are the sum of a known external part and an unknown constraint part. Hence we write the governing equations of p^α and $\mathscr{B}^\beta$ as

$$m^\alpha \frac{d^2x^{\alpha,i}}{dt^2} = f_e^{\alpha,i} + f_c^{\alpha,i}, \tag{5.1}$$

$$M^\beta \frac{d^2\xi^{\beta,i}}{dt^2} = F^{\beta,i} = F_e^{\beta,i} + F_c^{\beta,i}, \tag{5.2}$$

$$I^{\beta,ij}W^{\beta,jk}\frac{d^2\theta^{\beta,k}}{dt^2} + \left(I^{\beta,ij}\frac{\partial W^{\beta,jk}}{\partial\theta^{\beta,l}} + \varepsilon^{ijr}W^{\beta,jl}I^{\beta,rs}W^{\beta,sk}\right)\frac{d\theta^{\beta,l}}{dt}\frac{d\theta^{\beta,k}}{dt}$$
$$= G^{\beta,i} = G_e^{\beta,i} + G_c^{\beta,i}, \tag{5.3}$$

where $\alpha = 1, \ldots, n$, $\beta = 1, \ldots, K$, and $i = 1, 2, 3$ are free indices, while j, k, l, r, s are dummy indices summed from 1 to 3. The subscripts e and c of $f^{\alpha,i}$, $F^{\beta,i}$, and $G^{\beta,i}$ denote the external part and the constraint part, respectively. We shall do away with these cumbersome notations as soon as we have obtained the governing equations for the whole system $\mathscr{Q}$.

Now we multiply (5.1) by $\partial\varphi^{\alpha,i}/\partial q^{\Delta}$, (5.2) by $\partial\psi^{\beta,i}/\partial q^{\Delta}$, and (5.3) by $W^{\beta,ih}\,\partial\zeta^{\beta,h}/\partial q^{\Delta}$, and then sum the results with respect to all repeated indices over their appropriate ranges, obtaining

$$
\begin{aligned}
m^{\alpha}\frac{\partial\varphi^{\alpha,i}}{\partial q^{\Delta}}\frac{d^2x^{\alpha,i}}{dt^2}&+M^{\beta}\frac{\partial\psi^{\beta,i}}{\partial q^{\Delta}}\frac{d^2\xi^{\beta,i}}{dt^2}+I^{\beta,ij}W^{\beta,ih}W^{\beta,jk}\frac{\partial\zeta^{\beta,h}}{\partial q^{\Delta}}\frac{d^2\theta^{\beta,k}}{dt^2}\\
&+\left(I^{\beta,ij}\frac{\partial W^{\beta,jk}}{\partial\theta^{\beta,l}}+\varepsilon^{ijr}W^{\beta,jl}I^{\beta,rs}W^{\beta,sk}\right)W^{\beta,ih}\frac{\partial\zeta^{\beta,h}}{\partial q^{\Delta}}\frac{d\theta^{\beta,l}}{dt}\frac{d\theta^{\beta,k}}{dt}\\
&=\frac{\partial\varphi^{\alpha,i}}{\partial q^{\Delta}}f_e^{\alpha,i}+\frac{\partial\psi^{\beta,i}}{\partial q^{\Delta}}F_e^{\beta,i}+W^{\beta,ih}\frac{\partial\zeta^{\beta,h}}{\partial q^{\Delta}}G_e^{\beta,i}+\frac{\partial\varphi^{\alpha,i}}{\partial q^{\Delta}}f_c^{\alpha,i}\\
&+\frac{\partial\psi^{\beta,i}}{\partial q^{\Delta}}F_c^{\beta,i}+W^{\beta,ih}\frac{\partial\zeta^{\beta,h}}{\partial q^{\Delta}}G_c^{\beta,i},
\end{aligned}
\tag{5.4}
$$

where $\Delta=1,\ldots,n$ is the only free index. Since $x^{\alpha,i}$, $\xi^{\beta,i}$, $\theta^{\beta,i}$ can all be expressed in terms of the generalized coordinates q^{Δ} through the constraint relations (3.24), by using the chain rule we can regard (5.4) as a system of second-order differential equations in $q^{\Delta}(t)$. These equations will determine $q^{\Delta}(t)$ as functions of t provided that the right-hand side be a certain known function of t.

Notice that the right-hand side of (5.4) is the sum of an external part and a constraint part. We now assume that the constraint part in (5.4) vanishes; i.e.,

$$
\frac{\partial\varphi^{\alpha,i}}{\partial q^{\Delta}}f_c^{\alpha,i}+\frac{\partial\psi^{\beta,i}}{\partial q^{\Delta}}F_c^{\beta,i}+W^{\beta,ih}\frac{\partial\zeta^{\beta,h}}{\partial q^{\Delta}}G_c^{\beta,i}=0 \tag{5.5}
$$

for each $\Delta=1,\ldots,n$. Under this assumption, the right-hand side of (5.4) reduces to the external part only and is generally a certain known function.

It should be noted that the assumption (5.5) does not require that the constraint parts $f_c^{\alpha,i}$, $F_c^{\beta,i}$, $G_c^{\beta,i}$ of the force systems on the individual elements p^{α}, $\mathscr{B}^{\beta}$ of $\mathscr{Q}$ be equal to zero separately. Indeed, there are $3N+6K$ components of constraint forces and moments in (5.1)–(5.3) but there are only n equations in the system (5.5). Since the coefficient matrix of (5.5) has rank n (this condition corresponds to the linear independence of the tangent vectors $\{\mathbf{h}_{\Delta}\}$ of the q^{Δ} coordinate curves as explained in Section 3), the vanishing of the sum in (5.5) implies the vanishing of the components $f_c^{\alpha,i}$, $F_c^{\beta,i}$, $G_c^{\beta,i}$ if and only if $n=3N+6K$. In that case, the constraint surface $\mathscr{M}$ is just the free configuration space, $\mathscr{R}^{3N}\times\mathscr{R}^{3K}\times\mathscr{SO}(3)^K$, as it should be, since the vanishing of the individual constraint forces means literally that the system is free.

When $n < 3N + 6K$, the assumption (5.5) may be interpreted in the following way: For a fixed time t_0 and position $\mathbf{q}_0 \in \mathscr{M}^{t_0}$ we call any smooth curve $\mathbf{q}(\tau) \in \mathscr{M}^{t_0}$ passing through $\mathbf{q}_0$ a *virtual displacement* from $\mathbf{q}_0$. Since $\mathbf{q}(\tau)$ is required to stay in $\mathscr{M}^{t_0}$, we say that the virtual displacement is *consistent with the constraint*. We claim that the condition (5.5) is satisfied at $\mathbf{q} = \mathbf{q}_0$ and at $t = t_0$ if and only if the total power of the constraint forces and the constraint moments vanishes identically at the point $\mathbf{q}_0$ in all virtual displacements from $\mathbf{q}_0$. The truth of this claim is more or less obvious. Indeed, the power $P_c(\tau)$ of the constraint forces and the constraint moments in any virtual displacement $\mathbf{q}(\tau)$ is given by

$$P_c = f_c^{\alpha,i} \frac{dx^{\alpha,i}}{d\tau} + F_c^{\beta,i} \frac{d\xi^{\beta,i}}{d\tau} + G_c^{\beta,i} W^{\beta,ih} \frac{d\theta^{\beta,h}}{d\tau}. \tag{5.6}$$

But since the virtual displacement $q^\Delta = q^\Delta(\tau)$ belongs to $\mathscr{M}^{t_0}$, from (3.23) we have

$$\begin{aligned} x^{\alpha,i}(\tau) &= \varphi^{\alpha,i}(q^\Delta(\tau), t_0), \\ \xi^{\beta,i}(\tau) &= \varphi^{\beta,i}(q^\Delta(\tau), t_0), \\ \theta^{\beta,i}(\tau) &= \zeta^{\beta,i}(q^\Delta(\tau), t_0). \end{aligned} \tag{5.7}$$

Substituting (5.7) into (5.6) and evaluating the result at $\mathbf{q}_0$, we get

$$P_c \,|_{\mathbf{q}_0} = \left(f_c^{\alpha,i} \frac{\partial \varphi^{\alpha,i}}{\partial q^\Delta} + F_c^{\beta,i} \frac{\partial \psi^{\beta,i}}{\partial q^\Delta} + G_c^{\beta,i} W^{\beta,ih} \frac{\partial \zeta^{\beta,h}}{\partial q^\Delta} \right) \Bigg|_{(\mathbf{q}_0, t_0)} \frac{dq^\Delta}{d\tau} \Bigg|_{\mathbf{q}_0}. \tag{5.8}$$

Consequently $P_c \,|_{\mathbf{q}_0} = 0$ for all directions $(dq^\Delta/d\tau)\,|_{\mathbf{q}_0}$ if and only if (5.5) holds at $(\mathbf{q}_0, t_0)$. This result shows that the assumption (5.5) asserts that the *constraint surface* $\mathscr{M}^{t_0}$ *is frictionless*. Under this assumption, the right-hand side of (5.4) is determined entirely by the external forces and moments.

We define

$$Q_\Delta \equiv f_e^{\alpha,i} \frac{\partial \varphi^{\alpha,i}}{\partial q^\Delta} + F_e^{\beta,i} \frac{\partial \psi^{\beta,i}}{\partial q^\Delta} + G_e^{\beta,i} W^{\beta,ih} \frac{\partial \zeta^{\beta,h}}{\partial q^\Delta}, \qquad \Delta = 1, \ldots, n. \tag{5.9}$$

The significance of these quantities can be seen easily from the power P_e of the external forces and moments in the virtual displacement considered before. In fact, by using the argument leading to (5.8), we now have

$$\begin{aligned} P_e \,|_{\mathbf{q}_0} &= \left(f_e^{\alpha,i} \frac{\partial \varphi^{\alpha,i}}{\partial q^\Delta} + F_e^{\beta,i} \frac{\partial \psi^{\beta,i}}{\partial q^\Delta} + G_e^{\beta,i} W^{\beta,ih} \frac{\partial \zeta^{\beta,h}}{\partial q^\Delta} \right) \Bigg|_{(\mathbf{q}_0, t_0)} \frac{dq^\Delta}{d\tau} \Bigg|_{\mathbf{q}_0} \\ &= Q_\Delta \,|_{(\mathbf{q}_0, t_0)} \frac{dq^\Delta}{d\tau} \Bigg|_{\mathbf{q}_0}. \end{aligned} \tag{5.10}$$

Since $(d\mathbf{q}^{\Delta}/d\tau)\,|_{\mathbf{q}_0}$ are the components of the generalized velocity of the virtual displacement in $\mathscr{M}^{t_0}$ at $\mathbf{q}_0$, the formula (5.10) requires Q_Δ to be the components of a *generalized force*. Indeed, from (5.9) we can prove easily that Q_Δ satisfy the transformation law of the components of a covariant vector under a change of generalized coordinates on the instantaneous constraint surface $\mathscr{M}^{t_0}$. As before let $(\bar{q}^{\Delta})$ be related to (q^{Δ}) by (3.19). Then at the fixed time t_0 the surface $\mathscr{M}^{t_0}$ can be represented by

$$x^{\alpha,i} = \bar{\varphi}^{\alpha,i}(\bar{q}^{\Delta}, t_0), \qquad \xi^{\beta,i} = \bar{\psi}^{\beta,i}(\bar{q}^{\Delta}, t_0), \qquad \theta^{\beta,i} = \bar{\zeta}^{\beta,i}(\bar{q}^{\Delta}, t_0) \tag{5.11}$$

also. Substituting these into (5.9) and using the chain rule based on (3.19), we obtain the desired transformation law:

$$\begin{aligned}\bar{Q}_\Delta &= f_e^{\alpha,i}\frac{\partial\bar{\varphi}^{\alpha,i}}{\partial\bar{q}^{\Delta}} + F_e^{\beta,i}\frac{\partial\bar{\psi}^{\beta,i}}{\partial\bar{q}^{\Delta}} + G_e^{\beta,i}W^{\beta,ih}\frac{\partial\bar{\zeta}^{\beta,h}}{\partial\bar{q}^{\Delta}} \\ &= \left(f_e^{\alpha,i}\frac{\partial\varphi^{\alpha,i}}{\partial q^{\Gamma}} + F_e^{\beta,i}\frac{\partial\psi^{\beta,i}}{\partial q^{\Gamma}} + G_e^{\beta,i}W^{\beta,ih}\frac{\partial\zeta^{\beta,h}}{\partial q^{\Gamma}}\right)\frac{\partial q^{\Gamma}}{\partial\bar{q}^{\Delta}} \\ &= Q_\Gamma\frac{\partial q^{\Gamma}}{\partial\bar{q}^{\Delta}}, \qquad \Delta = 1, \ldots, n.\end{aligned} \tag{5.12}$$

Since the power P_e is taken relative to a virtual displacement in $\mathscr{M}^{t_0}$ only, the quantities Q_Δ are the tangential components of the generalized force in the instantaneous constraint surface. The actual power P of the motion is taken along the trajectory (3.3) of the system and is given by

$$P = f^{\alpha,i}\frac{dx^{\alpha,i}}{dt} + F^{\beta,i}\frac{d\xi^{\beta,i}}{dt} + G^{\beta,i}W^{\beta,ih}\frac{d\theta^{\beta,h}}{dt}, \tag{5.13}$$

where the total forces and moments $\mathbf{f}$, $\mathbf{F}$, $\mathbf{G}$ are used. Substituting (3.28)–(3.30) into (5.13) and using the assumption (5.5), we obtain

$$P = Q_\Delta\frac{dq^{\Delta}}{dt} + f^{\alpha,i}\frac{\partial\varphi^{\alpha,i}}{\partial t} + F^{\beta,i}\frac{\partial\psi^{\beta,i}}{\partial t} + G^{\beta,i}W^{\beta,ih}\frac{\partial\zeta^{\beta,h}}{\partial t}. \tag{5.14}$$

Comparing this result with the component form (3.27) of the generalized velocity $\dot{\mathbf{q}}$, we see that the last three terms on the right-hand side of (5.14) correspond to the component of the generalized force associated with the basis vector $\mathbf{h}_{n+1}$. Thus we put

$$Q_{n+1} = f^{\alpha,i}\frac{\partial\varphi^{\alpha,i}}{\partial t} + F^{\beta,i}\frac{\partial\psi^{\beta,i}}{\partial t} + G^{\beta,i}W^{\beta,ih}\frac{\partial\zeta^{\beta,h}}{\partial t}. \tag{5.15}$$

Relative to the dual basis $\{\mathbf{h}^{\Delta}, \mathbf{h}^{n+1}, \Delta = 1, \ldots, n\}$ of $\{\mathbf{h}_{\Delta}, \mathbf{h}_{n+1}, \Delta = 1, \ldots, n\}$, the generalized force $\mathbf{Q}$ is given by the component form

$$\mathbf{Q} = Q_{\Delta}\mathbf{h}^{\Delta} + Q_{n+1}\mathbf{h}^{n+1}. \tag{5.16}$$

The power P is, then, simply the value of $\mathbf{Q}$ at the generalized velocity $\dot{\mathbf{q}}$,

$$P = \langle \mathbf{Q}, \dot{\mathbf{q}} \rangle = Q_{\Delta}\frac{dq^{\Delta}}{dt} + Q_{n+1}. \tag{5.17}$$

It should be noted that both the external and the constraint forces and moments enter into the component Q_{n+1} of $\mathbf{Q}$. This fact is not hard to understand, since when the constraint surface is moving in the free configuration space, the constraint forces and moments, though having no tangential components, still contribute toward the power of the system.

So far we have considered the right-hand side of the system (5.4). Now we turn our attention to the left-hand side. We claim that it may be rewritten as

$$\frac{d}{dt}\left(\frac{\partial E}{\partial \dot{q}^{\Delta}}\right) - \frac{\partial E}{\partial q^{\Delta}}, \tag{5.18}$$

where

$$E = E(q^1, \ldots, q^n, \dot{q}^1, \ldots, \dot{q}^n, t)$$

denotes the kinetic energy of the system $\mathscr{Q}$. We shall prove this result by expanding (5.18) and showing that it coincides with the left-hand side of (5.4).

Recall first that for the general case of time-dependent constraint the kinetic energy E is given by the formula (3.16) with the summation on Δ and Γ from 1 to $n + 1$. We can, of course, substitute (3.16) into (5.18) and calculate the result directly. But this approach is very complicated, since we have to differentiate E twice in the expression (5.18). A somewhat simpler way of showing that (5.18) coincides with the left-hand side of (5.4) is as follows: We start from the original definition of the kinetic energy function:

$$E = \tfrac{1}{2}(m^{\alpha}\dot{x}^{\alpha,i}\dot{x}^{\alpha,i} + M^{\beta}\dot{\xi}^{\beta,i}\dot{\xi}^{\beta,i} + W^{\beta,ij}I^{\beta,ik}W^{\beta,kl}\dot{\theta}^{\beta,j}\dot{\theta}^{\beta,l}), \tag{5.19}$$

where the repeated indices $\alpha, \beta, i, j, k, l$ are summed over their appropriate ranges: α, 1 to N; β, 1 to K; i, j, k, l, 1 to 3. As we have shown by (3.16), E is really a function of q^{Δ}, $\dot{q}^{\Delta}$, and t. The dependence of E on these variables can be obtained from that of $\dot{x}^{\alpha,i}$, $\dot{\xi}^{\beta,i}$, $\dot{\theta}^{\beta,i}$, and $W^{\beta,ij}$. Specifically, from

(3.28), (3.29), (3.30), and (2.27), respectively, we have

$$\dot{x}^{\alpha,i} = \dot{x}^{\alpha,i}(q^\Delta, \dot{q}^\Delta, t) = \frac{\partial \varphi^{\alpha,i}}{\partial q^\Gamma} \dot{q}^\Gamma + \frac{\partial \varphi^{\alpha,i}}{\partial t}, \tag{5.20}$$

$$\dot{\xi}^{\beta,i} = \dot{\xi}^{\beta,i}(q^\Delta, \dot{q}^\Delta, t) = \frac{\partial \psi^{\beta,i}}{\partial q^\Gamma} \dot{q}^\Gamma + \frac{\partial \psi^{\beta,i}}{\partial t}, \tag{5.21}$$

$$\dot{\theta}^{\beta,i} = \dot{\theta}^{\beta,i}(q^\Delta, \dot{q}^\Delta, t) = \frac{\partial \zeta^{\beta,i}}{\partial q^\Gamma} \dot{q}^\Gamma + \frac{\partial \zeta^{\beta,i}}{\partial t}, \tag{5.22}$$

and

$$[W^{\beta,ij}] = [W^{\beta,ij}(q^\Delta, t)] = \begin{bmatrix} -\sin \zeta^{\beta,2} \cos \zeta^{\beta,3} & \sin \zeta^{\beta,3} & 0 \\ \sin \zeta^{\beta,2} \sin \zeta^{\beta,3} & \cos \zeta^{\beta,3} & 0 \\ \cos \zeta^{\beta,2} & 0 & 1 \end{bmatrix}, \tag{5.23}$$

where $\varphi^{\alpha,i}$, $\psi^{\beta,i}$, $\zeta^{\beta,i}$ are functions of (q^Δ, t) given by the constraint relations (3.24). Using (5.19) and (5.20)–(5.23), we can now calculate the expression (5.18) by the chain rule.

First, taking the partial derivative of (5.19) with respect to $\dot{q}^\Delta$, we obtain

$$\frac{\partial E}{\partial \dot{q}^\Delta} = m^\alpha \frac{\partial \varphi^{\alpha,i}}{\partial q^\Delta} \dot{x}^{\alpha,i} + M^\beta \frac{\partial \psi^{\beta,i}}{\partial q^\Delta} \dot{\xi}^{\beta,i} + W^{\beta,ij} I^{\beta,ik} W^{\beta,kl} \frac{\partial \zeta^{\beta,j}}{\partial q^\Delta} \dot{\theta}^{\beta,l}, \tag{5.24}$$

where we have used the chain rule together with the following results of (5.20), (5.21), and (5.22):

$$\frac{\partial \dot{x}^{\alpha,i}}{\partial \dot{q}^\Delta} = \frac{\partial \varphi^{\alpha,i}}{\partial q^\Delta}, \qquad \frac{\partial \dot{\xi}^{\beta,i}}{\partial \dot{q}^\Delta} = \frac{\partial \psi^{\beta,i}}{\partial q^\Delta}, \qquad \frac{\partial \dot{\theta}^{\beta,i}}{\partial \dot{q}^\Delta} = \frac{\partial \zeta^{\beta,i}}{\partial q^\Delta}. \tag{5.25}$$

Now taking the time derivative of (5.24), we obtain

$$\begin{aligned} \frac{d}{dt}\left(\frac{\partial E}{\partial \dot{q}^\Delta}\right) &= m^\alpha \frac{\partial \varphi^{\alpha,i}}{\partial q^\Delta} \ddot{x}^{\alpha,i} + M^\beta \frac{\partial \psi^{\beta,i}}{\partial q^\Delta} \ddot{\xi}^{\beta,i} + W^{\beta,ij} I^{\beta,ik} W^{\beta,kl} \frac{\partial \zeta^{\beta,j}}{\partial q^\Delta} \ddot{\theta}^{\beta,l} \\ &\quad + m^\alpha \frac{\partial \dot{x}^{\alpha,i}}{\partial q^\Delta} \dot{x}^{\alpha,i} + M^\beta \frac{\partial \dot{\xi}^{\beta,i}}{\partial q^\Delta} \dot{\xi}^{\beta,i} \\ &\quad + W^{\beta,ij} I^{\beta,ik} W^{\beta,kl} \frac{\partial \dot{\theta}^{\beta,j}}{\partial q^\Delta} \dot{\theta}^{\beta,l} \\ &\quad + \frac{\partial W^{\beta,ij}}{\partial \theta^{\beta,r}} \dot{\theta}^{\beta,r} I^{\beta,ik} W^{\beta,kl} \frac{\partial \zeta^{\beta,j}}{\partial q^\Delta} \dot{\theta}^{\beta,l} \\ &\quad + W^{\beta,ij} I^{\beta,ik} \frac{\partial W^{\beta,kl}}{\partial \theta^{\beta,r}} \dot{\theta}^{\beta,r} \frac{\partial \zeta^{\beta,j}}{\partial q^\Delta} \dot{\theta}^{\beta,l}, \end{aligned} \tag{5.26}$$

where we have used the following results:

$$\frac{d}{dt}\left(\frac{\partial \varphi^{\alpha,i}}{\partial q^{\Delta}}\right) = \frac{\partial \dot{x}^{\alpha,i}}{\partial q^{\Delta}}, \quad \frac{d}{dt}\left(\frac{\partial \psi^{\beta,i}}{\partial q^{\Delta}}\right) = \frac{\partial \dot{\xi}^{\beta,i}}{\partial q^{\Delta}}, \quad \frac{d}{dt}\left(\frac{\partial \zeta^{\beta,i}}{\partial q^{\Delta}}\right) = \frac{\partial \dot{\theta}^{\beta,i}}{\partial q^{\Delta}}, \tag{5.27}$$

which can be derived directly from (5.20), (5.21), and (5.22).

Next, taking the partial derivative of (5.19) with respect to q^{Δ}, we obtain

$$\begin{aligned}\frac{\partial E}{\partial q^{\Delta}} &= m^{\alpha}\frac{\partial \dot{x}^{\alpha,i}}{\partial q^{\Delta}}\dot{x}^{\alpha,i} + M^{\beta}\frac{\partial \dot{\xi}^{\beta,i}}{\partial q^{\Delta}}\dot{\xi}^{\beta,i} + W^{\beta,ij}I^{\beta,ik}W^{\beta,kl}\frac{\partial \dot{\theta}^{\beta,j}}{\partial q^{\Delta}}\dot{\theta}^{\beta,l} \\ &\quad + \frac{\partial W^{\beta,ij}}{\partial \theta^{\beta,r}}\frac{\partial \zeta^{\beta,r}}{\partial q^{\Delta}}I^{\beta,ik}W^{\beta,kl}\dot{\theta}^{\beta,j}\dot{\theta}^{\beta,l}.\end{aligned} \tag{5.28}$$

Subtracting (5.28) from (5.26), we finally obtain

$$\begin{aligned}\frac{d}{dt}\left(\frac{\partial E}{\partial \dot{q}^{\Delta}}\right) - \frac{\partial E}{\partial q^{\Delta}} &= m^{\alpha}\frac{\partial \varphi^{\alpha,i}}{\partial q^{\Delta}}\ddot{x}^{\alpha,i} + M^{\beta}\frac{\partial \psi^{\beta,i}}{\partial q^{\Delta}}\ddot{\xi}^{\beta,i} \\ &\quad + W^{\beta,ij}I^{\beta,ik}W^{\beta,kl}\frac{\partial \zeta^{\beta,j}}{\partial q^{\Delta}}\ddot{\theta}^{\beta,l} \\ &\quad + W^{\beta,ih}I^{\beta,ij}\frac{\partial W^{\beta,jk}}{\partial \theta^{\beta,l}}\dot{\theta}^{\beta,l}\frac{\partial \zeta^{\beta,h}}{\partial q^{\Delta}}\dot{\theta}^{\beta,k} \\ &\quad + \left(\frac{\partial W^{\beta,ij}}{\partial \theta^{\beta,r}} - \frac{\partial W^{\beta,ir}}{\partial \theta^{\beta,j}}\right)I^{\beta,ik}W^{\beta,kl}\frac{\partial \zeta^{\beta,j}}{\partial q^{\Delta}}\dot{\theta}^{\beta,l}\dot{\theta}^{\beta,r},\end{aligned} \tag{5.29}$$

where we have changed several dummy indices in order to make the first four terms on the right-hand side of (5.29) appear exactly the same as the first four terms on the left-hand side of (5.4).

Now from (5.23) it can be proved by direct differentiation that the matrix $[W^{\beta,ij}]$ obeys the following identities:

$$\frac{\partial W^{\beta,ij}}{\partial \theta^{\beta,r}} - \frac{\partial W^{\beta,ir}}{\partial \theta^{\beta,j}} = \varepsilon^{spi}W^{\beta,sj}W^{\beta,pr} \qquad (\beta \text{ not summed}). \tag{5.30}$$

These identities are rather interesting. While the roles of the Eulerian angles $(\theta^{\beta,i})$ are quite different from one another—indeed, $[W^{\beta,ij}]$ does not depend on $\theta^{\beta,1}$ at all, and the dependence of $[W^{\beta,ij}]$ on $\theta^{\beta,2}$ and $\theta^{\beta,3}$ is not symmetrical—the identities (5.30), nevertheless, are symmetrical with respect to the free indices i, j, r, which may take on any values of 1, 2, 3. Using (5.30), we see immediately that the last term on the right-hand side of (5.29)—after the rearrangements of the dummy indices: s to i, p to j, i to r, j to k, r to l, and k to s—is exactly the same as the last term on the left-hand side of (5.4). Thus the proof is complete.

The results obtained so far can now be summarized by rewriting (5.4) as

$$\frac{d}{dt}\left(\frac{\partial E}{\partial \dot{q}^{\Delta}}\right) - \frac{\partial E}{\partial q^{\Delta}} = Q_{\Delta}, \qquad \Delta = 1, \ldots, n, \tag{5.31}$$

which are *Lagrange's equations* for the constrained system $\mathscr{Q}$. Since E is a quadratic function in $\dot{q}^{\Delta}$ with coefficients depending on q^{Δ} and t, (5.31) is a system of second-order differential equations for $q^{\Delta}(t)$. Assuming that the generalized forces Q_{Δ} on the right-hand side of (5.31) are known, we can solve the system and determine $q^{\Delta}(t)$ as functions of t provided that certain initial values, say $q^{\Delta}(o)$ and $\dot{q}^{\Delta}(o)$, are given. The standard coordinates $x^{\alpha,i}$, $\xi^{\beta,i}$, $\theta^{\beta,i}$ of the system can then be determined from $q^{\Delta}(t)$ by the constraint relations (3.2) or (3.24).

We shall see certain explicit forms of Lagrange's equations in the following section.

6. Explicit Forms of Lagrange's Equations

In this section we derive some explicit forms of Lagrange's equations for the special case when the constraint is time independent. As explained in Section 3, the kinetic energy E is given by a homogeneous quadratic function of $\dot{q}^{\Delta}$ in this case; cf. (3.16), where $[g_{\Delta\Gamma}]$ depends only on q^{Δ} and is independent of t. We have established the fact that $[g_{\Delta\Gamma}]$ is the component matrix of a Riemannian metric $\mathbf{g}$ on the constraint manifold $\mathscr{M}$. We shall now examine the meaning of Lagrange's equations (5.31) in the context of the Riemannian geometry on $\mathscr{M}$.

From (3.16) we can calculate the partial derivatives

$$\frac{\partial E}{\partial \dot{q}^{\Delta}} = g_{\Delta\Gamma}\dot{q}^{\Gamma}, \tag{6.1a}$$

$$\frac{\partial E}{\partial q^{\Delta}} = \frac{1}{2}\,\frac{\partial g_{\Gamma\Lambda}}{\partial q^{\Delta}}\,\dot{q}^{\Gamma}\dot{q}^{\Lambda}. \tag{6.1b}$$

Taking the time derivative of (6.1a) by the chain rule, we obtain

$$\frac{d}{dt}\left(\frac{\partial E}{\partial \dot{q}^{\Delta}}\right) = g_{\Delta\Gamma}\ddot{q}^{\Gamma} + \frac{\partial g_{\Delta\Gamma}}{\partial q^{\Lambda}}\,\dot{q}^{\Gamma}\dot{q}^{\Lambda}. \tag{6.2}$$

Substituting (6.1b) and (6.2) into (5.31), we get

$$g_{\Delta\Gamma}\ddot{q}^{\Gamma} + \left(\frac{\partial g_{\Delta\Gamma}}{\partial q^{\Lambda}} - \frac{1}{2}\frac{\partial g_{\Gamma\Lambda}}{\partial q^{\Delta}}\right)\dot{q}^{\Gamma}\dot{q}^{\Lambda} = Q_{\Delta}, \tag{6.3}$$

which is an explicit form of the system of Lagrange's equations. From (6.3) we see clearly that Lagrange's equations form a system of second-order differential equations for the generalized coordinates q^{Δ}.

Since the product $\dot{q}^{\Gamma}\dot{q}^{\Lambda}$ is symmetric with respect to the indices Γ and Λ, (6.3) is equivalent to

$$g_{\Delta\Gamma}\ddot{q}^{\Gamma} + \frac{1}{2}\left(\frac{\partial g_{\Delta\Gamma}}{\partial q^{\Lambda}} + \frac{\partial g_{\Delta\Lambda}}{\partial q^{\Gamma}} - \frac{\partial g_{\Gamma\Lambda}}{\partial q^{\Delta}}\right)\dot{q}^{\Gamma}\dot{q}^{\Lambda} = Q_{\Delta}. \tag{6.4}$$

This system is familiar in Riemannian geometry. Indeed, comparing (6.4) with (56.3) in Section 56, IVT-2, we see that (6.4) can be rewritten as

$$g_{\Delta\Gamma}\left(\ddot{q}^{\Gamma} + \begin{Bmatrix}\Gamma \\ \Lambda\Xi\end{Bmatrix}\dot{q}^{\Lambda}\dot{q}^{\Xi}\right) = Q_{\Delta}, \tag{6.5}$$

where the Christoffel symbols $\{{}^{\Gamma}_{\Lambda\Xi}\}$ are based on the metric $\mathbf{g}$, viz.,

$$\begin{Bmatrix}\Gamma \\ \Lambda\Xi\end{Bmatrix} = \frac{1}{2}g^{\Gamma\Sigma}\left(\frac{\partial g_{\Sigma\Lambda}}{\partial q^{\Xi}} + \frac{\partial g_{\Sigma\Xi}}{\partial q^{\Lambda}} - \frac{\partial g_{\Lambda\Xi}}{\partial q^{\Sigma}}\right), \tag{6.6}$$

and where $[g^{\Gamma\Xi}]$ is the inverse of $[g_{\Gamma\Sigma}]$, viz.,

$$g^{\Gamma\Sigma}g_{\Sigma\Delta} = \delta_{\Delta}{}^{\Gamma}. \tag{6.7}$$

Multiplying (6.5) by the inverse metric $g^{\Gamma\Sigma}$, we obtain

$$\ddot{q}^{\Gamma} + \begin{Bmatrix}\Gamma \\ \Lambda\Xi\end{Bmatrix}\dot{q}^{\Lambda}\dot{q}^{\Xi} = Q^{\Gamma}, \tag{6.8}$$

where Q^{Γ} denotes the contravariant components of the generalized force,

$$Q^{\Gamma} = g^{\Gamma\Delta}Q_{\Delta}. \tag{6.9}$$

The operation of raising the indices used in this equation is explained in detail in Section 35, IVT-1.

As explained in Section 56, IVT-2 [cf. equation (56.21)], the left-hand side of (6.8) is just the covariant derivative of the generalized velocity $\dot{\mathbf{q}} = \dot{q}^{\Delta}\mathbf{h}_{\Delta}$ along the trajectory $\mathbf{q} = \mathbf{q}(t)$. As a result, (6.8) can be written in the following coordinate-free form:

$$\frac{D\dot{\mathbf{q}}}{Dt} = \mathbf{Q}. \tag{6.10}$$

We call the covariant derivative $D\dot{\mathbf{q}}/Dt$ the *generalized acceleration*. Relative to any coordinate system (q^Δ) in $\mathscr{M}$ the contravariant component forms of the generalized velocity and the generalized acceleration are

$$\dot{\mathbf{q}} = \dot{q}^\Delta \mathbf{h}_\Delta, \qquad \frac{D\dot{\mathbf{q}}}{Dt} = \left(\ddot{q}^\Delta + \begin{Bmatrix} \Delta \\ \Gamma\Lambda \end{Bmatrix} \dot{q}^\Gamma \dot{q}^\Lambda\right)\mathbf{h}_\Delta. \tag{6.11}$$

Lagrange's equations (6.3) or (6.4) simply express (6.10) in component form and may be interpreted as the assertion that the result of the inertia metric applied to the generalized acceleration is equal to the generalized force.

In particular, when the system consists of a single particle p and is free of any constraint, Lagrange's equations (6.5) reduce to Newton's equations (6.3), which require that the acceleration multiplied by the mass be equal to the force, since the inertia metric is simply the physical metric multiplied by the mass in this case. When the system consists in a single rigid body $\mathscr{B}$ and is free of any constraint, Lagrange's equations (6.5) are the combination of the six equations (4.11) and (4.12). In this case the inertia metric $\mathbf{g}$ on the free configuration space $\mathscr{R}^3 \times \mathscr{SO}(3)$ is the direct sum of the product of M with the standard metric on $\mathscr{R}^3$ and the transformed inertia metric $W^{ij}I^{ik}W^{kl}$ on $\mathscr{SO}(3)$; cf. (5.19). Consequently, the first three components of Lagrange's equations are the equations (4.12). The classical Euler's equations (4.9) or (4.10), however, are not Lagrange's equations relative to any coordinate system on $\mathscr{SO}(3)$; they correspond to Lagrange's equations relative to a certain anholonomic basis, as we now explain.

In Section 46, IVT-2, we have defined the concept of an anholonomic basis: a smooth field of bases which may fail to satisfy the integrability condition of a coordinate basis (i.e., it need not be the natural basis field of any coordinate system). On the constraint manifold $\mathscr{M}$, let $\{\mathbf{k}_\Delta\}$ be an anholonomic basis. To specify $\{\mathbf{k}_\Delta\}$, we use its component form relative to the natural basis $\{\mathbf{h}_\Delta\}$ of the coordinate system (q^Δ):

$$\mathbf{k}_\Delta = k_\Delta{}^\Gamma \mathbf{h}_\Gamma, \qquad \mathbf{h}_\Delta = h_\Delta{}^\Gamma \mathbf{k}_\Gamma. \tag{6.12}$$

Then the condition that $\{\mathbf{k}_\Delta\}$ fails to be integrable may be written as follows:

$$\mathbf{0} \neq [\mathbf{k}_\Delta, \mathbf{k}_\Gamma] = \left(\frac{\partial k_\Gamma{}^\Lambda}{\partial q^\Sigma} k_\Delta{}^\Sigma - \frac{\partial k_\Delta{}^\Lambda}{\partial q^\Sigma} k_\Gamma{}^\Sigma \right) h_\Lambda{}^\Xi \mathbf{k}_\Xi \equiv K^\Xi_{\Delta\Gamma} \mathbf{k}_\Xi. \tag{6.13}$$

Since $\{\mathbf{k}_\Delta\}$ is a field of bases on $\mathscr{M}$, we can express the vector equation (6.10) in component form relative to $\{\mathbf{k}_\Delta\}$. We proceed to work out the details.

First, the generalized velocity $\dot{\mathbf{q}}$ may be expressed in component form relative to $\{\mathbf{k}_\Delta\}$:

$$\dot{\mathbf{q}} = \omega^\Delta \mathbf{k}_\Delta, \tag{6.14}$$

where the anholonomic components ω^Δ are related to the holonomic components $\dot{q}^\Delta$ by the usual transformation law:

$$\dot{q}^\Delta = k_\Gamma{}^\Delta \omega^\Gamma, \qquad \omega^\Gamma = h_\Delta{}^\Gamma \dot{q}^\Delta. \tag{6.15}$$

Then the kinetic energy E may be regarded as a quadratic function of ω^Δ with coefficients depending on q^Δ, viz.,

$$\begin{aligned} E = E(q^\Delta, \dot{q}^\Delta) &= \tfrac{1}{2} g_{\Delta\Gamma}(q^\Delta) \dot{q}^\Delta \dot{q}^\Gamma, \\ &= \bar{E}(q^\Delta, \omega^\Delta) = \tfrac{1}{2} \bar{g}_{\Delta\Gamma}(q^\Delta) \omega^\Delta \omega^\Gamma, \end{aligned} \tag{6.16}$$

where $\bar{g}_{\Delta\Gamma}$ are the anholonomic components of $\mathbf{g}$ relative to $\{\mathbf{k}_\Delta\}$ and are related to $g_{\Delta\Gamma}$ by the transformation law:

$$\bar{g}_{\Delta\Gamma} = k_\Delta{}^\Lambda k_\Gamma{}^\Sigma g_{\Lambda\Sigma}. \tag{6.17}$$

From (6.16) the partial derivatives $\partial E/\partial \dot{q}^\Delta$ and $\partial \bar{E}/\partial \omega^\Delta$ are related by

$$\frac{\partial E}{\partial \dot{q}^\Delta} = g_{\Delta\Gamma} \dot{q}^\Gamma = g_{\Delta\Gamma} k_\Lambda{}^\Gamma \omega^\Lambda = \frac{\partial \bar{E}}{\partial \omega^\Lambda} h_\Delta{}^\Gamma. \tag{6.18}$$

Similarly, the partial derivatives $\partial E/\partial q^\Delta$ and $\partial \bar{E}/\partial q^\Delta$ are related by

$$\frac{\partial E}{\partial q^\Delta} = \frac{\partial \bar{E}}{\partial q^\Delta} + \frac{\partial \bar{E}}{\partial \omega^\Gamma} \dot{q}^\Lambda \frac{\partial h_\Lambda{}^\Gamma}{\partial q^\Delta} = \frac{\partial \bar{E}}{\partial q^\Delta} + \frac{\partial \bar{E}}{\partial \omega^\Gamma} \omega^\Xi k_\Xi{}^\Lambda \frac{\partial h_\Lambda{}^\Gamma}{\partial q^\Delta}. \tag{6.19}$$

As usual we denote the dual basis of $\{\mathbf{k}_\Delta\}$ by $\{\mathbf{k}^\Delta\}$. Then the generalized force $\mathbf{Q}$ may be expressed in component form relative to $\{\mathbf{k}^\Delta\}$,

$$\mathbf{Q} = Q_\Gamma \mathbf{h}^\Gamma = \Omega_\Gamma \mathbf{k}^\Gamma, \tag{6.20}$$

where Q_Γ and Ω_Γ are related by the transformation law:

$$Q_\Gamma = \Omega_\Delta h_\Gamma{}^\Delta, \qquad \Omega_\Delta = Q_\Gamma k_\Delta{}^\Gamma. \tag{6.21}$$

From (6.15), (6.21), and (5.10) the anholonomic components Ω_Δ of the generalized force $\mathbf{Q}$ are simply the coefficients of the linear function

$$P = Q_\Delta \dot{q}^\Delta = \Omega_\Delta \omega^\Delta, \tag{6.22}$$

which gives the power in any virtual displacement consistent with the constraint.

Now substituting (6.18), (6.19), and (6.21) into (5.31), we obtain

$$\frac{d}{dt}\left(h_\Delta{}^\Gamma \frac{\partial \bar{E}}{\partial \omega^\Gamma}\right) - \frac{\partial \bar{E}}{\partial q^\Delta} - \frac{\partial \bar{E}}{\partial \omega^\Gamma} \omega^\Xi k_\Xi{}^\Lambda \frac{\partial h_\Lambda{}^\Gamma}{\partial q^\Delta} = h_\Delta{}^\Gamma \Omega_\Gamma. \tag{6.23}$$

We can expand the leading term into a more familiar form by using (6.15) and the product rule

$$\frac{d}{dt}\left(h_\Delta{}^\Gamma \frac{\partial \bar{E}}{\partial \omega^\Gamma}\right) = h_\Delta{}^\Gamma \frac{d}{dt}\left(\frac{\partial \bar{E}}{\partial \omega^\Gamma}\right) + \frac{\partial \bar{E}}{\partial \omega^\Gamma} \frac{\partial h_\Delta{}^\Gamma}{\partial q^\Lambda} \omega^\Xi k_\Xi{}^\Lambda. \tag{6.24}$$

Then (6.23) can be rewritten as

$$\frac{d}{dt}\left(\frac{\partial \bar{E}}{\partial \omega^\Gamma}\right) - k_\Gamma{}^\Delta \frac{\partial \bar{E}}{\partial q^\Delta} + \frac{\partial \bar{E}}{\partial \omega^\Sigma} \omega^\Xi k_\Xi{}^\Lambda k_\Gamma{}^\Delta \left(\frac{\partial h_\Delta{}^\Sigma}{\partial q^\Lambda} - \frac{\partial h_\Lambda{}^\Sigma}{\partial q^\Delta}\right) = \Omega_\Gamma, \tag{6.25}$$

where we have multiplied the equation by $[k_\Gamma{}^\Delta]$, which is the inverse of $[h_\Delta{}^\Gamma]$. Differentiating the identity

$$h_\Delta{}^\Sigma k_\Gamma{}^\Delta = \delta_\Gamma{}^\Sigma \tag{6.26}$$

with respect to q^Λ, we obtain

$$\frac{\partial h_\Delta{}^\Sigma}{\partial q^\Lambda} = -h_\Delta{}^\Gamma h_\Xi{}^\Sigma \frac{\partial k_\Gamma{}^\Xi}{\partial q^\Lambda}. \tag{6.27}$$

Hence (6.25) is equivalent to

$$\frac{d}{dt}\left(\frac{\partial \bar{E}}{\partial \omega^\Gamma}\right) - k_\Gamma{}^\Delta \frac{\partial \bar{E}}{\partial q^\Delta} + \frac{\partial \bar{E}}{\partial \omega^\Sigma} \omega^\Xi h_\Lambda{}^\Sigma \left(\frac{\partial k_\Xi{}^\Lambda}{\partial q^\Delta} k_\Gamma{}^\Delta - \frac{\partial k_\Gamma{}^\Lambda}{\partial q^\Delta} k_\Xi{}^\Delta\right) = \Omega_\Gamma. \tag{6.28}$$

Comparing the third term on the left-hand side with the condition of integrability (6.13), we can rewrite (6.28) as

$$\frac{d}{dt}\left(\frac{\partial \bar{E}}{\partial \omega^{\Gamma}}\right) - k_{\Gamma}{}^{\Delta}\frac{\partial \bar{E}}{\partial q^{\Delta}} + \frac{\partial \bar{E}}{\partial \omega^{\Sigma}}\,\omega^{\Xi}K_{\Gamma\Xi}^{\Sigma} = \Omega_{\Gamma}, \tag{6.29}$$

which is the system of Lagrange's equations relative to the anholonomic basis $\{\mathbf{k}_{\Delta}\}$.

It should be noted that, if the basis $\{\mathbf{k}_{\Delta}\}$ is, in fact, the natural basis of a coordinate system $(\bar{q}^{\Delta})$, then ω^{Γ} is just $\dot{\bar{q}}^{\Gamma}$, and $[k_{\Gamma}{}^{\Delta}]$ is just the Jacobian matrix $[\partial q^{\Delta}/\partial \bar{q}^{\Gamma}]$ of the coordinate transformation. In this case $K_{\Gamma\Xi}^{\Sigma}$ vanish identically, of course, since $\{\mathbf{k}_{\Delta}\}$, being a coordinate basis, satisfies the condition of integrability, $[\mathbf{k}_{\Delta}, \mathbf{k}_{\Gamma}] = \mathbf{0}$. As a result (6.29) reduces to

$$\frac{d}{dt}\left(\frac{\partial \bar{E}}{\partial \dot{\bar{q}}^{\Gamma}}\right) - \frac{\partial \bar{E}}{\partial \bar{q}^{\Gamma}} = \bar{Q}_{\Gamma}, \tag{6.30}$$

which is just the system of Lagrange's equations in the $(\bar{q}^{\Delta})$ coordinate system.

Next we show that Euler's equations (4.9) may be regarded as a special case of (6.29). We consider a single rigid body $\mathscr{B}$, which is free of any constraints. On the free configuration space $\mathscr{R}^3 \times \mathscr{SO}(3)$ we choose the anholonomic basis $\{\mathbf{e}_i, \mathbf{k}_i\}$, where $\{\mathbf{e}_i\}$ is the standard basis on $\mathscr{R}^3$, but $\{\mathbf{k}_i\}$ is an anholonomic basis with component forms

$$\mathbf{h}_i = W^{ij}\mathbf{k}_j, \qquad \mathbf{k}_i = (W^{-1})^{ij}\mathbf{h}_j, \tag{6.31}$$

relative to the natural basis $\{\mathbf{h}_i\}$ of the coordinate system (θ^i). We choose $[W^{ij}]$ to be the particular coefficient matrix in (2.27), so that the anholonomic components ω^i of the generalized velocity $\dot{\boldsymbol{\theta}}$ are precisely the components of the angular velocity relative to the imbedded frame. Further, from (6.13), (6.31), and (5.30) we have

$$[\mathbf{k}_i, \mathbf{k}_j] = \varepsilon^{ijl}\mathbf{k}_l \neq 0, \tag{6.32}$$

i.e., in this case we know that $\{\mathbf{k}_i\}$ fails to be a holonomic basis.

From (2.40) the kinetic energy E is given by

$$E = \bar{E}(\xi^i, \theta^i, \dot{\xi}^i, \omega^i) = \tfrac{1}{2}M\dot{\xi}^i\dot{\xi}^i + \tfrac{1}{2}I^{ij}\omega^i\omega^j \tag{6.33}$$

in terms of the anholonomic components ω^i. Hence the partial derivatives of $\bar{E}$ with respect to ω^i and θ^i are

$$\frac{\partial \bar{E}}{\partial \omega^i} = I^{ij}\omega^j, \qquad \frac{\partial \bar{E}}{\partial \theta^i} = 0. \tag{6.34}$$

In terms of the angular velocity $\boldsymbol{\omega}$, the power P is given by

$$P = F^i \dot{\xi}^i + G^i \omega^i. \tag{6.35}$$

Hence the anholonomic components of the generalized force relative to the basis $\{\mathbf{k}_i\}$ are precisely the components G^i of the moment relative to the imbedded frame at the center of mass. Using the results (6.32), (6.34), and (6.35), we obtain from (6.29) the system

$$\frac{d}{dt}(I^{ij}\omega^j) + I^{kl}\omega^l\omega^j\varepsilon^{ijk} = G^i, \tag{6.36}$$

which is just the system of Euler's equations (4.9).

The preceding analysis shows that the Newtonian equations and Euler's equations may be recovered from Lagrange's equations. Thus no mathematical information is lost when we pass from the Newtonian equations and Euler's equations to Lagrange's equations. We have gained the information that the generalized forces become known functions, however, when the constraint forces and the constraint moments obey the assumption (5.5). This new information is the main advantage of Lagrange's equations over the Newtonian equations and Euler's equations.

2

Hamiltonian Systems in Phase Space

The equations of Lagrange, which we have derived in Chapter 1, may be transformed into a system of first-order differential equations when the generalized force possesses a potential function. The transformation is known as the *Legendre transformation* and the resulting first-order system is known as the *Hamiltonian system*. In this chapter we develop the theory of Hamiltonian systems from the standpoint of a flow problem and in the context of a certain first-order partial differential equation known as the *Hamilton–Jacobi equation*.

7. Hamilton's Principle

In the preceding chapter we showed that the motions of a constrained dynamical system $\mathscr{Q}$ are governed by Lagrange's equations:

$$\frac{d}{dt}\left(\frac{\partial E}{\partial \dot{q}^{\Delta}}\right) - \frac{\partial E}{\partial q^{\Delta}} = Q_{\Delta}, \qquad \Delta = 1, \ldots, n. \tag{7.1}$$

We can solve these equations and determine q^{Δ} as functions of t provided that Q_{Δ} are certain known functions of t. In this section we wish to consider a slightly different problem. We assume that Q_{Δ} are given implicitly by a generalized potential function $V = V(q^{\Delta}, \dot{q}^{\Delta}, t)$ in the following way:

$$Q_{\Delta} = \frac{d}{dt}\left(\frac{\partial V}{\partial \dot{q}^{\Delta}}\right) - \frac{\partial V}{\partial q^{\Delta}}, \tag{7.2}$$

where the time derivative is taken along the trajectory of the system. Under this assumption (7.1) may be rewritten as

$$\frac{d}{dt}\left(\frac{\partial L}{\partial \dot{q}^{\Delta}}\right) - \frac{\partial L}{\partial q^{\Delta}} = 0, \qquad \Delta = 1, \ldots, n, \tag{7.3}$$

where $L = L(q^{\Delta}, \dot{q}^{\Delta}, t)$ is defined by

$$L = E(q^{\Delta}, \dot{q}^{\Delta}, t) - V(q^{\Delta}, \dot{q}^{\Delta}, t) \tag{7.4}$$

and is called the *kinetic potential* or the *Lagrangian function* of the system.

An important special case of (7.2) is the case when V is a function of q^{Δ} and t only,

$$V = V(q^{\Delta}, t), \tag{7.5}$$

so that (7.2) reduces to

$$Q_{\Delta} = -\frac{\partial V}{\partial q^{\Delta}}. \tag{7.6}$$

For this special case the generalized force $\mathbf{Q}$ is simply the gradient of the potential function V on each instantaneous constraint surface $\mathscr{M}^t$. In particular, the power in any virtual displacement $q^{\Delta} = q^{\Delta}(\tau)$ in $\mathscr{M}^t$ is given by

$$P = Q_{\Delta}\frac{dq^{\Delta}}{d\tau} = -\frac{\partial V}{\partial q^{\Delta}}\frac{dq^{\Delta}}{d\tau} = -\frac{d}{d\tau}V(q^{\Delta}(\tau), t), \tag{7.7}$$

where t is held fixed. The work done in this virtual displacement from $q^{\Delta}(0)$ to $q^{\Delta}(\tau)$ is then given by

$$\int_0^{\tau} P(\tau)\,d\tau = V(q^{\Delta}(0), t) - V(q^{\Delta}(\tau), t), \tag{7.8}$$

which depends only on the terminal points $(q^{\Delta}(0), t)$ and $(q^{\Delta}(\tau), t)$ of the virtual displacement. For this reason $V(q^{\Delta}, t)$ may be regarded as the virtual potential energy on $\mathscr{M}^t$. When V depends on t, (7.8) is valid only for a virtual displacement in $\mathscr{M}^t$. For the actual trajectory $q^{\Delta} = q^{\Delta}(t)$ of $\mathscr{Q}$, the power $P(t)$ is given by

$$P = Q_{\Delta}\frac{dq^{\Delta}}{dt} = -\frac{\partial V}{\partial q^{\Delta}}\frac{dq^{\Delta}}{dt} = -\frac{d}{dt}V(q^{\Delta}(t), t) + \frac{\partial V}{\partial t}, \tag{7.9}$$

which contains an extra term $\partial V/\partial t$ due to the explicit dependence of V on time.

The main difference between the special case (7.6) and the general case (7.2) is that in the special case Q_Δ depends only on q^Δ and t, while in the general case Q_Δ depends on q^Δ, t, as well as on the time derivatives $\dot{q}^\Delta$ and $\ddot{q}^\Delta$. Regardless of which of these two cases is used, the Lagrangian function L defined by (7.4) generally depends on $\dot{q}^\Delta$, since the kinetic energy E is a quadratic function of $\dot{q}^\Delta$. In this section we are primarily interested in the system (7.3), where L is a function in general of the variables q^Δ, $\dot{q}^\Delta$, and t. This function need not be quadratic in $\dot{q}^\Delta$, since the potential function V may depend arbitrarily on $(q^\Delta, \dot{q}^\Delta, t)$.

The system (7.3) has a familiar form. In Section 57, IVT-2, we have remarked that (7.3) is the system of Euler–Lagrange equations for the time integral of the Lagrangian function L. In the context of analytical mechanics this assertion is known as *Hamilton's principle*. Specifically, we consider the integral

$$I\big(q^\Delta(t), t \in [a, b]\big) = \int_a^b L\big(q^\Delta(t), \dot{q}^\Delta(t), t\big)\, dt. \tag{7.10}$$

We claim that the trajectory $q^\Delta = q^\Delta(t)$ satisfies the system (7.3) if and only if it is an extremal curve for the integral I in the following sense: Consider any 1-parameter family of curves $q^\Delta = q_\varepsilon{}^\Delta(t)$ of the form

$$q_\varepsilon{}^\Delta(t) = q^\Delta(t) + \varepsilon\eta^\Delta(t), \qquad t \in [a, b], \tag{7.11}$$

where η^Δ satisfies the conditions

$$\eta^\Delta(a) = \eta^\Delta(b) = 0, \qquad \Delta = 1, \ldots, n, \tag{7.12}$$

so that the curves $\mathbf{q}_\varepsilon$ all have the same terminal points: $q_\varepsilon{}^\Delta(a) = q^\Delta(a)$, $q_\varepsilon{}^\Delta(b) = q^\Delta(b)$ for all ε. Then Hamilton's principle asserts that

$$\frac{d}{d\varepsilon} I\big(q_\varepsilon{}^\Delta(t), t \in [a, b]\big)\big|_{\varepsilon=0} = 0 \tag{7.13}$$

for all choice of η^Δ if and only if $q^\Delta(t) = q_0{}^\Delta(t)$ satisfies the system (7.3).

We can prove Hamilton's principle by a standard argument in the calculus of variations. First, taking the time derivative of (7.11), we obtain

$$\dot{q}_\varepsilon{}^\Delta(t) = \dot{q}^\Delta(t) + \varepsilon\dot{\eta}^\Delta(t). \tag{7.14}$$

Substituting (7.11) and (7.14) into (7.10), we get

$$I\big(q_\varepsilon{}^\Delta(t), t \in [a, b]\big) = \int_a^b L\big(q^\Delta(t) + \varepsilon\eta^\Delta(t), \dot{q}^\Delta(t) + \varepsilon\dot{\eta}^\Delta(t), t\big)\, dt. \tag{7.15}$$

Since the limits a and b of the integral are independent of the parameter ε, we can differentiate (7.15) with respect to ε under the integral sign,

$$\frac{d}{d\varepsilon} I(q_\varepsilon{}^\Delta(t), t \in [a, b]) = \int_a^b \left(\frac{\partial L}{\partial q^\Delta} \eta^\Delta + \frac{\partial L}{\partial \dot{q}^\Delta} \dot{\eta}^\Delta\right) dt. \tag{7.16}$$

The second term in the integrand may be integrated by parts, yielding

$$\begin{aligned} \int_a^b \frac{\partial L}{\partial \dot{q}^\Delta} \dot{\eta}^\Delta \, dt &= \frac{\partial L}{\partial \dot{q}^\Delta} \eta^\Delta \Big|_a^b - \int_a^b \frac{d}{dt}\left(\frac{\partial L}{\partial \dot{q}^\Delta}\right)\eta^\Delta \, dt \\ &= 0 - \int_a^b \frac{d}{dt}\left(\frac{\partial L}{\partial \dot{q}^\Delta}\right)\eta^\Delta \, dt, \end{aligned} \tag{7.17}$$

where we have used the conditions (7.12). Combining (7.16) and (7.17), and evaluating the result at $\varepsilon = 0$, we get

$$\frac{d}{d\varepsilon} I(q_\varepsilon{}^\Delta(t), t \in [a, b]) \,\Big|_{\varepsilon=0} = \int_a^b \left[\frac{\partial L}{\partial q^\Delta} - \frac{d}{dt}\left(\frac{\partial L}{\partial \dot{q}^\Delta}\right)\right]\Bigg|_{\varepsilon=0} \eta^\Delta \, dt. \tag{7.18}$$

Hamilton's principle now follows directly from this equation. Indeed, since the functions $\eta^\Delta(t)$ are arbitrary, the integral vanishes if and only if the coefficients of η^Δ in the integrand all vanish; i.e., (7.3) holds.

Before closing this section we mention that an important special case of the system (7.3) is the case when the Lagrangian function L is independent of the time t. We can characterize this special case by the following result: The Lagrangian L is independent of t if and only if the system (7.3) possesses an integral of the form

$$H = \frac{\partial L}{\partial \dot{q}^\Delta} \dot{q}^\Delta - L, \tag{7.19}$$

which is known as the *Jacobi integral*. To prove this assertion, we simply take the time derivative of H and evaluate the result on any trajectory of (7.3):

$$\begin{aligned} \frac{d}{dt} H &= \frac{d}{dt}\left(\frac{\partial L}{\partial \dot{q}^\Delta} \dot{q}^\Delta - L\right) \\ &= \frac{d}{dt}\left(\frac{\partial L}{\partial \dot{q}^\Delta}\right)\dot{q}^\Delta + \frac{\partial L}{\partial \dot{q}^\Delta} \ddot{q}^\Delta - \frac{\partial L}{\partial q^\Delta} \dot{q}^\Delta - \frac{\partial L}{\partial \dot{q}^\Delta} \ddot{q}^\Delta - \frac{\partial L}{\partial t} \\ &= \left[\frac{d}{dt}\left(\frac{\partial L}{\partial \dot{q}^\Delta}\right) - \frac{\partial L}{\partial q^\Delta}\right]\dot{q}^\Delta - \frac{\partial L}{\partial t} = 0 - \frac{\partial L}{\partial t}. \end{aligned} \tag{7.20}$$

Thus H is constant on each trajectory if and only if L is independent of t.

It should be noted that the condition $\partial L/\partial t = 0$ is naturally satisfied when the constraint and the potential function are both independent of t. In this case the Jacobi integral is simply the total energy of the system. Indeed, when the constraint is time independent, the kinetic energy E is given by a homogeneous quadratic function of $\dot{q}^\Delta$ as shown in (3.16). From Euler's theorem, we then have

$$\frac{\partial L}{\partial \dot{q}^\Delta}\dot{q}^\Delta = \frac{\partial E}{\partial \dot{q}^\Delta}\dot{q}^\Delta = 2E, \tag{7.21}$$

where we have assumed that $V = V(q^\Delta)$. Combining (7.21) and (7.19), we get

$$H = 2E - E + V = E + V, \tag{7.22}$$

so that the Jacobi integral corresponds to the usual law of conservation of total energy on each trajectory.

When the constraint is time independent, the system (7.3) can be rewritten as a system of autonomous first-order differential equations

$$\frac{dq^\Delta}{dt} = v^\Delta, \qquad \frac{\partial^2 L}{\partial v^\Delta\, \partial v^\Gamma}\frac{dv^\Gamma}{dt} = \frac{\partial L}{\partial q^\Delta} - \frac{\partial^2 L}{\partial q^\Gamma\, \partial v^\Delta}v^\Gamma, \tag{7.23}$$

where (q^Δ, v^Δ) are the dependent variables, and where L is regarded as a function of q^Δ and v^Δ. We shall assume that the Hessian matrix $[\partial^2 L/\partial v^\Delta \partial v^\Gamma]$ of L with respect to v^Δ is nonsingular. Then the first-order time derivatives dq^Δ/dt, dv^Δ/dt of q^Δ and v^Δ are given by certain functions of q^Γ and v^Γ independent of t.

We can regard the first-order system (7.23) as the system of governing equations for the flow generated by a certain vector field on the manifold $\mathscr{T}(\mathscr{M})$ formed by the pairs $(\mathbf{q}, \mathbf{v})$, where $\mathbf{q}$ is a point in the constraint manifold $\mathscr{M}$, and where $\mathbf{v}$ is a vector belonging to the tangent space $\mathscr{M}_\mathbf{q}$ of $\mathscr{M}$ at $\mathbf{q}$. Specifically, we define $\mathscr{T}(\mathscr{M})$ by

$$\mathscr{T}(\mathscr{M}) = \bigcup_{\mathbf{q}\in\mathscr{M}} \mathscr{M}_\mathbf{q}. \tag{7.24}$$

It is a manifold of dimension $2n$, called the *tangent bundle* of $\mathscr{M}$. Like the manifold $\mathscr{M}$, which is a surface of dimension n in a Euclidean space $\mathscr{R}^{3N}\times\mathscr{R}^{3K}\times\mathscr{R}^{9K}$, the manifold $\mathscr{T}(\mathscr{M})$ can be regarded as a surface of dimension $2n$ in the Euclidean space $(\mathscr{R}^{3N}\times\mathscr{R}^{3K}\times\mathscr{R}^{9K})^2$. Indeed, the tangent space $\mathscr{M}_\mathbf{q}$ at any point $\mathbf{q}\in\mathscr{M}$ is a subspace of dimension n in the translation space $\mathscr{R}^{3N}\times\mathscr{R}^{3K}\times\mathscr{R}^{9K}$ of the underlying Euclidean space

$\mathscr{R}^{3N} \times \mathscr{R}^{3K} \times \mathscr{R}^{9K}$. Naturally, a coordinate system for $\mathscr{T}(\mathscr{M})$ is given by (q^Δ, v^Δ), where (q^Δ) is a coordinate system for $\mathscr{M}$, and where (v^Δ) is the Cartesian coordinate system on $\mathscr{M}_\mathbf{q}$ induced by the natural basis $\{\mathbf{h}_\Delta(\mathbf{q})\}$ of (q^Δ) at $\mathbf{q} \in \mathscr{M}$.

In Section 49, IVT-2, we have considered the problem of the flow generated by a vector field in general. The first-order system (7.23) is just a special case of that problem for a vector field on the manifold $\mathscr{T}(\mathscr{M})$. It turns out that the system (7.23) may be analyzed more effectively by first transforming it into a system on another manifold, $\mathscr{T}^*(\mathscr{M})$, called the *cotangent bundle* or the *phase space* of $\mathscr{M}$. This transformation is called the *Legendre transformation* and is based on the assumption that the Hessian matrix $[\partial^2 L/\partial v^\Delta \partial v^\Gamma]$ in (7.23) is nonsingular. The result of the Legendre transformation is known as the *Hamiltonian system*, which is a system of first-order differential equations on $\mathscr{T}^*(\mathscr{M})$ having a very special form. We shall discuss these topics in detail in the subsequent sections of this chapter.

8. Phase Space and Its Canonical Differential Forms

In this section we consider the special case that the constraint is time independent; i.e., the constraint surface $\mathscr{M}$ is a fixed manifold of dimension n in the free configuration space. In Section 3 we have defined the tangent space $\mathscr{M}_\mathbf{q}$ of $\mathscr{M}$ at $\mathbf{q} \in \mathscr{M}$. We have pointed out that $\mathscr{M}_\mathbf{q}$ is an n-dimensional vector space which depends on the point $\mathbf{q}$. Relative to any coordinate system (q^Δ) $\mathscr{M}_\mathbf{q}$ is spanned by the natural basis $\{\mathbf{h}_\Delta(\mathbf{q})\}$, where $\mathbf{h}_\Delta(\mathbf{q})$ denotes the tangent vector of the coordinate curve of q^Δ at $\mathbf{q}$.

Suppose that $\mathbf{q} = \mathbf{q}(t)$ is a curve in $\mathscr{M}$ with coordinates $q^\Delta = q^\Delta(t)$. Then the tangent vector $\dot{\mathbf{q}}$ is given by the component form

$$\dot{\mathbf{q}} = \dot{q}^\Delta \mathbf{h}_\Delta \tag{8.1}$$

relative to the natural basis $\{\mathbf{h}_\Delta\}$. In particular, under any change of coordinates

$$\bar{q}^\Delta = \bar{q}^\Delta(q^1, \ldots, q^n), \qquad q^\Delta = q^\Delta(\bar{q}^1, \ldots, \bar{q}^n), \tag{8.2}$$

the tangent vectors $\bar{\mathbf{h}}_\Delta$ of the $\bar{q}^\Delta$ coordinate curves are given by the component form

$$\bar{\mathbf{h}}_\Delta = \frac{\partial q^\Gamma}{\partial \bar{q}^\Delta} \mathbf{h}_\Gamma, \qquad \Delta = 1, \ldots, n. \tag{8.3}$$

We can regard (8.3) as a change of basis in $\mathscr{M}_{\mathbf{q}}$ for any $\mathbf{q} \in \mathscr{M}$. Then the components v^{Δ} and $\bar{v}^{\Delta}$ of any tangent vector $\mathbf{v} \in \mathscr{M}_{\mathbf{q}}$ relative to the natural bases $\{\mathbf{h}_{\Delta}(\mathbf{q})\}$ and $\{\bar{\mathbf{h}}_{\Delta}(\mathbf{q})\}$ are related by the system of transformation laws

$$v^{\Delta} = \frac{\partial q^{\Delta}}{\partial \bar{q}^{\Gamma}}\bigg|_{\mathbf{q}} \bar{v}^{\Gamma}, \qquad \bar{v}^{\Delta} = \frac{\partial \bar{q}^{\Delta}}{\partial q^{\Gamma}}\bigg|_{\mathbf{q}} v^{\Gamma}. \tag{8.4}$$

For this reason a tangent vector corresponds to a *contravariant vector*.

As usual we denote the dual space of $\mathscr{M}_{\mathbf{q}}$ by $\mathscr{M}_{\mathbf{q}}^*$, and we call $\mathscr{M}_{\mathbf{q}}^*$ the *cotangent space* of $\mathscr{M}$ at $\mathbf{q}$. A basis for $\mathscr{M}_{\mathbf{q}}^*$ is the dual basis $\{\mathbf{h}^{\Delta}(\mathbf{q})\}$ of $\{\mathbf{h}_{\Delta}(\mathbf{q})\}$. As explained in Section 31, IVT-1, the components of any cotangent vector $\mathbf{p} \in \mathscr{M}_{\mathbf{q}}^*$ relative to the dual basis $\{\mathbf{h}^{\Delta}(\mathbf{q})\}$ obey the system of transformation laws

$$p_{\Delta} = \frac{\partial \bar{q}^{\Gamma}}{\partial q^{\Delta}}\bigg|_{\mathbf{q}} \bar{p}_{\Gamma}, \tag{8.5a}$$

$$\bar{p}_{\Delta} = \frac{\partial q^{\Gamma}}{\partial \bar{q}^{\Delta}}\bigg|_{\mathbf{q}} p_{\Gamma}, \tag{8.5b}$$

which is the dual of (8.4). As a result, a cotangent vector corresponds to a *covariant vector*. From (8.1) we see that any $\mathbf{v} \in \mathscr{M}_{\mathbf{q}}$ can be regarded as the tangent vector of a certain curve at $\mathbf{q}$. The dual of this interpretation for $\mathscr{M}_{\mathbf{q}}^*$ is the assertion that any $p \in \mathscr{M}_{\mathbf{q}}^*$ can be regarded as the *gradient* or the *differential* of a certain function at $\mathbf{q}$.

The preceding assertion can be explained in the following way: Given any function f on $\mathscr{M}$, the *gradient* or the *differential* $d_{\mathbf{q}}f$ of f at $\mathbf{q}$ is the linear map $d_{\mathbf{q}}f: \mathscr{M}_{\mathbf{q}} \to \mathscr{R}$ such that

$$[d_{\mathbf{q}}f](\dot{\mathbf{q}}) = \frac{d}{dt} f(q(t))\,|_{\mathbf{q}} \tag{8.6}$$

for any curve $\mathbf{q} = \mathbf{q}(t)$ passing through $\mathbf{q}$. To show that $d_{\mathbf{q}}f$ is well defined by the condition (8.6), we must verify that the right-hand side of (8.6) depends linearly on the tangent vector $\dot{\mathbf{q}}$. This fact is more or less obvious, since by the chain rule we have

$$\frac{d}{dt} f(\mathbf{q}(t)) = \frac{\partial f}{\partial q^{\Delta}} \frac{dq^{\Delta}}{dt}. \tag{8.7}$$

Comparing this result with the component form (8.1), we see that $d_{\mathbf{q}}f \in \mathscr{M}_{\mathbf{q}}^*$, and that the component form for $d_{\mathbf{q}}f$ is

$$d_{\mathbf{q}}f = \frac{\partial f}{\partial q^{\Delta}}\bigg|_{\mathbf{q}} \mathbf{h}^{\Delta}(\mathbf{q}). \tag{8.8}$$

Thus the cotangent space $\mathscr{M}_{\mathbf{q}}{}^*$ is formed by the gradients of functions at $\mathbf{q}$. From (8.8) the dual basis $\{\mathbf{h}^\Delta(\mathbf{q})\}$ is formed by the gradients $\{d_{\mathbf{q}}q^\Delta\}$ of the coordinate functions (q^Δ) at $\mathbf{q}$.

At the end of the preceding section we have mentioned that the tangent bundle $\mathscr{T}(\mathscr{M})$, defined by (7.24), is a manifold of dimension $2n$. Now we define the *cotangent bundle* or the *phase space* $\mathscr{T}^*(\mathscr{M})$ by

$$\mathscr{T}^*(\mathscr{M}) = \bigcup_{\mathbf{q}\in\mathscr{M}} \mathscr{M}_{\mathbf{q}}{}^*. \tag{8.9}$$

As remarked before, $\mathscr{T}^*(\mathscr{M})$ is a manifold of dimension $2n$ and can be regarded as a surface in Euclidean space $(\mathscr{R}^{3N}\times\mathscr{R}^{3K}\times\mathscr{R}^{9K})^2$. A coordinate system for $\mathscr{T}^*(\mathscr{M})$ is given by (q^Δ, p_Δ), where (q^Δ) is a coordinate system for $\mathscr{M}$ as before, and where (p_Δ) is the Cartesian coordinate system on $\mathscr{M}_{\mathbf{q}}{}^*$ induced by the basis $\{\mathbf{h}^\Delta(\mathbf{q})\}$.

Now since $\mathscr{T}^*(\mathscr{M})$ is a manifold, it, too, has a tangent space $\mathscr{T}^*(\mathscr{M})_{(\mathbf{q},\mathbf{p})}$ and a cotangent space $\mathscr{T}^*(\mathscr{M})^*_{(\mathbf{q},\mathbf{p})}$ at each point $(\mathbf{q},\mathbf{p})\in\mathscr{T}^*(\mathscr{M})$. Using the coordinate system (q^Δ, p_Δ), we can define a basis $\{\mathbf{H}_\Delta, \mathbf{K}^\Delta, \Delta = 1,\ldots,n\}$ for $\mathscr{T}^*(\mathscr{M})_{(\mathbf{q},\mathbf{p})}$ and a basis $\{\mathbf{H}^\Delta, \mathbf{K}_\Delta, \Delta = 1,\ldots,n\}$ for $\mathscr{T}^*(\mathscr{M})^*_{(\mathbf{q},\mathbf{p})}$, where

$$\begin{aligned}
&\mathbf{H}_\Delta \text{ is the tangent vector of the } q^\Delta \text{ coordinate curve,}\\
&\mathbf{K}^\Delta \text{ is the tangent vector of the } p_\Delta \text{ coordinate curve,}\\
&\mathbf{H}^\Delta \text{ is the gradient of the } q^\Delta \text{ coordinate function,}\\
&\mathbf{K}_\Delta \text{ is the gradient of the } p_\Delta \text{ coordinate function.}
\end{aligned} \tag{8.10}$$

As before, $\{\mathbf{H}^\Delta, \mathbf{K}_\Delta\}$ is the dual basis of $\{\mathbf{H}_\Delta, \mathbf{K}^\Delta\}$. That is, $\langle\mathbf{H}^\Delta, \mathbf{H}_\Gamma\rangle = \langle\mathbf{K}_\Gamma, \mathbf{K}^\Delta\rangle = \delta_\Gamma{}^\Delta$, while $\langle\mathbf{H}^\Delta, \mathbf{K}^\Gamma\rangle = \langle\mathbf{K}_\Delta, \mathbf{H}_\Gamma\rangle = 0$ for all. $\Delta, \Gamma = 1, \ldots, n$. Here the $\langle\ ,\ \rangle$ operation is defined on the dual spaces $\mathscr{T}^*(\mathscr{M})_{(\mathbf{q},\mathbf{p})}$ and $\mathscr{T}^*(\mathscr{M})^*_{(\mathbf{q},\mathbf{p})}$.

In Sections 51 and 55, IVT-2, we have defined the concepts of differential forms and exterior derivative. A differential r-form is just a smooth field of skew-symmetric covariant tensors of order r, and the exterior derivative is an operator d, which maps an r-form to an $(r+1)$-form in accord with the formulas (51.5) or (55.34), IVT-2. We now apply these concepts to the cotangent bundle $\mathscr{T}^*(\mathscr{M})$.

First, $\mathscr{T}^*(\mathscr{M})$ is endowed with a canonical 1-form $\boldsymbol{\theta}$, which has the component form

$$\boldsymbol{\theta} = p_\Delta\mathbf{H}^\Delta = p_\Delta\, dq^\Delta. \tag{8.11}$$

Notice that this 1-form is a linear combination of the basis $\{\mathbf{H}^\Delta, \mathbf{K}_\Delta\}$

of (q^Δ, p_Δ) in $\mathscr{C}^*(\mathscr{M})$, the components of $\boldsymbol{\theta}$ in $\mathbf{H}^\Delta$ being p_Δ and the components in $\mathbf{K}_\Delta$ being 0. The reason for calling this 1-form *canonical* is because $\boldsymbol{\theta}$ is independent of the choice of coordinates (q^Δ, p_Δ). In other words, when we change the coordinates (q^Δ) to $(\bar{q}^\Delta)$, and we consider a corresponding change of coordinates from (q^Δ, p_Δ) to $(\bar{q}^\Delta, \bar{p}_\Delta)$ for $\mathscr{C}^*(\mathscr{M})$, the component form for $\boldsymbol{\theta}$ is still given by

$$\boldsymbol{\theta} = \bar{p}_\Delta \mathbf{H}^\Delta = \bar{p}_\Delta \, d\bar{q}^\Delta. \tag{8.12}$$

This fact is by no means obvious. We shall now explain the argument leading to it.

From the definition (8.9) for $\mathscr{C}^*(\mathscr{M})$ we see that there is a natural projection map $\boldsymbol{\pi}: \mathscr{C}^*(\mathscr{M}) \to \mathscr{M}$ such that

$$\boldsymbol{\pi}(\mathbf{q}, \mathbf{p}) = \mathbf{q} \tag{8.13}$$

for all $(\mathbf{q}, \mathbf{p}) \in \mathscr{C}^*(\mathscr{M})$. Using this mapping, we can transform a function f on $\mathscr{M}$ into a function $f \circ \boldsymbol{\pi}$ on $\mathscr{C}^*(\mathscr{M})$. (This transformation is just the trivial operation of regarding a function of $\mathbf{q}$ as a function of $\mathbf{q}$ and $\mathbf{p}$.) As a result, the gradient $d_\mathbf{q} f$ is transformed into the gradient $d_{(\mathbf{q},\mathbf{p})}(f \circ \boldsymbol{\pi})$ in such a way that the components $\partial f/\partial q^\Delta$ of $d_\mathbf{q} f$ relative to $\{\mathbf{h}^\Delta\}$ [cf. (8.8)] become the components of $d_{(\mathbf{q},\mathbf{p})}(f \circ \boldsymbol{\pi})$ but relative to $\{\mathbf{H}^\Delta, \mathbf{K}_\Delta\}$, viz.,

$$[\text{grad}\, \boldsymbol{\pi}](d_\mathbf{q} f) \equiv d_{(\mathbf{q},\mathbf{p})}(f \circ \boldsymbol{\pi}) = \left.\frac{\partial f}{\partial q^\Delta}\right|_\mathbf{q} \mathbf{H}^\Delta(\mathbf{q}, \mathbf{p}). \tag{8.14}$$

Now since any cotangent vector $\mathbf{p} \in \mathscr{M}_\mathbf{q}^*$ corresponds to the gradient of a certain function, say f,

$$\mathbf{p} = p_\Delta \mathbf{h}^\Delta(\mathbf{q}) = \left.\frac{\partial f}{\partial q^\Delta}\right|_\mathbf{q} \mathbf{h}^\Delta(\mathbf{q}), \tag{8.15}$$

by applying the transformation $\boldsymbol{\pi}$, we can transform the cotangent vector $\mathbf{p} = p_\Delta \mathbf{h}^\Delta(\mathbf{q}) \in \mathscr{M}_\mathbf{q}^*$ into the cotangent vector $p_\Delta \mathbf{H}^\Delta(\mathbf{q}, \mathbf{p}) \in \mathscr{C}^*(\mathscr{M})^*_{(\mathbf{q},\mathbf{p})}$, which is just the assigned value $\boldsymbol{\theta}(\mathbf{q}, \mathbf{p})$ of the 1-form $\boldsymbol{\theta}$ at the point $(\mathbf{q}, \mathbf{p}) \in \mathscr{C}^*(\mathscr{M})$ [cf. (8.11)]. As a result, the value of the 1-form $\boldsymbol{\theta}$ is determined directly at each point $(\mathbf{q}, \mathbf{p})$ by the pair $(\mathbf{q}, \mathbf{p})$, and thus $\boldsymbol{\theta}$ is independent of any coordinate system (q^Δ).

We can prove the invariance of (8.11), viz.,

$$p_\Delta \mathbf{H}^\Delta = \bar{p}_\Delta \mathbf{H}^\Delta, \tag{8.16}$$

by considering the transformation law of the natural basis under a change of coordinates from (q^Δ, p_Δ) to $(\bar{q}^\Delta, \bar{p}_\Delta)$ also. As usual (q^Δ) and $(\bar{q}^\Delta)$ are related by (8.2) and their corresponding (p_Δ) and $(\bar{p}_\Delta)$ are related by (8.5). Consequently, by the argument leading to the system of transformation laws (8.3), we now have the system of transformation laws:

$$\bar{\mathbf{H}}_\Delta = \frac{\partial q^\Gamma}{\partial \bar{q}^\Delta} \mathbf{H}_\Gamma + \frac{\partial^2 \bar{q}^\Xi}{\partial q^\Gamma \partial q^\Lambda} \frac{\partial q^\Lambda}{\partial \bar{q}^\Delta} \bar{p}_\Xi \mathbf{K}^\Gamma, \tag{8.17a}$$

$$\bar{\mathbf{K}}^\Delta = \frac{\partial \bar{q}^\Delta}{\partial q^\Gamma} \mathbf{K}^\Gamma, \tag{8.17b}$$

and

$$\bar{\mathbf{H}}^\Delta = \frac{\partial \bar{q}^\Delta}{\partial q^\Gamma} \mathbf{H}^\Gamma, \tag{8.18a}$$

$$\bar{\mathbf{K}}_\Delta = \frac{\partial^2 q^\Xi}{\partial \bar{q}^\Delta \partial \bar{q}^\Lambda} \frac{\partial \bar{q}^\Lambda}{\partial q^\Gamma} p_\Xi \mathbf{H}^\Gamma + \frac{\partial q^\Gamma}{\partial \bar{q}^\Delta} \mathbf{K}_\Gamma. \tag{8.18b}$$

Using (8.5b) and (8.18a), we obtain

$$\bar{p}_\Delta \bar{\mathbf{H}}^\Delta = \frac{\partial q^\Gamma}{\partial \bar{q}^\Delta} p_\Gamma \frac{\partial \bar{q}^\Delta}{\partial q^\Lambda} \mathbf{H}^\Lambda = p_\Gamma \delta_\Lambda{}^\Gamma \mathbf{H}^\Lambda = p_\Lambda \mathbf{H}^\Lambda. \tag{8.19}$$

Thus (8.16) is proved.

Next, we define a canonical 2-form[1] $\boldsymbol{\omega}$ on $\mathscr{T}^*(\mathscr{M})$ by taking the exterior derivative of the canonical 1-form $\boldsymbol{\theta}$,

$$\boldsymbol{\omega} \equiv d\boldsymbol{\theta}. \tag{8.20}$$

Since the exterior derivative is an operator which is independent of the choice of coordinate system (cf. Section 51, IVT-2), the 2-form $\boldsymbol{\omega}$ defined by (8.20) enjoys the property of being canonical on $\mathscr{T}^*(\mathscr{M})$ as the 1-form $\boldsymbol{\theta}$. From the component form (8.11) for $\boldsymbol{\theta}$ and the basic properties, (51.21) and (51.24), IVT-2, for the exterior derivative we can determine easily the component form

$$\boldsymbol{\omega} = dp_\Delta \wedge dq^\Delta = dp_\Delta \otimes dq^\Delta - dq^\Delta \otimes dp_\Delta = \mathbf{K}_\Delta \otimes \mathbf{H}^\Delta - \mathbf{H}^\Delta \otimes \mathbf{K}_\Delta, \tag{8.21}$$

where $\wedge$ denotes the wedge product or the exterior product, which is defined in Section 38, IVT-1. The component form (8.21) indicates that $\boldsymbol{\omega}$ is a skew-symmetric covariant tensor field of order 2 on the manifold $\mathscr{T}^*(\mathscr{M})$.

[1] This notation should not be confused with that of the angular velocity.

In tensor algebra (cf. Sections 32 and 33, IVT-1) we have explained that a covariant tensor of order 2 over a vector space corresponds to a bilinear function on the vector space and to a linear map from the vector space to its dual space. Here the vector space over which $\boldsymbol{\omega}(\mathbf{q}, \mathbf{p})$ is a covariant tensor of order 2 is the tangent space $\mathscr{T}^*(\mathscr{M})_{(\mathbf{p},\mathbf{q})}$. From (8.21) the covariant tensor $\boldsymbol{\omega}$ corresponds to the bilinear function ω such that

$$\begin{aligned} \omega(\mathbf{H}_\Delta, \mathbf{H}_\Gamma) &= 0, \qquad & \omega(\mathbf{H}_\Delta, \mathbf{K}^\Gamma) &= -\delta_\Delta{}^\Gamma, \\ \omega(\mathbf{K}^\Gamma, \mathbf{H}_\Delta) &= \delta_\Delta{}^\Gamma, \qquad & \omega(\mathbf{K}^\Delta, \mathbf{K}^\Gamma) &= 0, \end{aligned} \tag{8.22}$$

and to the linear map such that

$$\boldsymbol{\omega}(\mathbf{H}_\Delta) = \mathbf{K}_\Delta, \qquad \boldsymbol{\omega}(\mathbf{K}^\Delta) = -\mathbf{H}^\Delta. \tag{8.23}$$

From (8.22) we see that the bilinear function defined by $\boldsymbol{\omega}$ is skew symmetric, as it should be, and from (8.23) we see that the linear map defined by $\boldsymbol{\omega}$ is nonsingular, since it transforms the basis $\{\mathbf{H}_\Delta, \mathbf{K}^\Delta\}$ for the tangent space into the basis $\{\mathbf{H}^\Delta, \mathbf{K}_\Delta\}$ for the cotangent space on the manifold $\mathscr{T}^*(\mathscr{M})$.

Another important property of the canonical 2-form $\boldsymbol{\omega}$ is that, from its very definition (8.20), $\boldsymbol{\omega}$ is an exact 2-form. In particular, $\boldsymbol{\omega}$ is closed; i.e.,

$$d\boldsymbol{\omega} = \mathbf{0}. \tag{8.24}$$

[The concepts of exact forms and closed forms are defined in Section 52, IVT-2. The result (8.24) follows from the general property that $d^2 = \mathbf{0}$; cf. (51.24), IVT-2.]

Now a manifold which is equipped with a closed, nonsingular 2-form, such as the canonical 2-form $\boldsymbol{\omega}$ just considered, is called a *symplectic manifold*, and the particular 2-form is called its *symplectic form*. In this sense we have shown that the manifold $\mathscr{T}^*(\mathscr{M})$ is a symplectic manifold, and that the canonical 2-form $\boldsymbol{\omega}$ is its symplectic form. This symplectic structure on the phase space $\mathscr{T}^*(\mathscr{M})$ is important in the theory of the Hamiltonian system, which is the main topic of the following section.

9. The Legendre Transformation and the Hamiltonian System I: The Time-Independent Case

In Section 7 we remarked that the second-order system of Lagrange's equations (7.3) on the constraint manifold $\mathscr{M}$ can be replaced by the first-order system (7.23) on the tangent bundle $\mathscr{T}(\mathscr{M})$ of $\mathscr{M}$. We have pointed out that when the Lagrangian function $L(q^\Delta, v^\Delta)$ is regular (i.e., when the Hessian matrix $[\partial^2 L/\partial v^\Delta \partial v^\Gamma]$ is nonsingular), (7.23) is an autonomous system whose flow problem is equivalent to the dynamical problem of finding the trajectories of the constrained system. In this section we show that the flow problem of (7.23) can be analyzed in an effective way by first transforming (7.23) into a system on the phase space $\mathscr{T}^*(\mathscr{M})$. This transformation is a change of variables from $(\mathbf{q}, \mathbf{v}) \in \mathscr{T}(\mathscr{M})$ to $(\mathbf{q}, \mathbf{p}) \in \mathscr{T}^*(\mathscr{M})$ defined by the condition

$$\mathbf{p}(\mathbf{u}) \equiv \frac{d}{ds} L(\mathbf{q}, \mathbf{v} + s\mathbf{u}) \big|_{s=0} \tag{9.1}$$

for all $\mathbf{u} \in \mathscr{M}_\mathbf{q}$. We call this change of variables the *Legendre transformation.*

To show that the Legendre transformation is well defined by the condition (9.1), we have to verify that the right-hand side of (9.1) is a linear function of $\mathbf{u}$ at each fixed $(\mathbf{q}, \mathbf{v}) \in \mathscr{T}(\mathscr{M})$. This fact follows directly from the chain rule:

$$\frac{d}{ds} L(\mathbf{q}, \mathbf{v} + s\mathbf{u}) \big|_{s=0} = \frac{\partial L}{\partial v^\Delta}\bigg|_{(\mathbf{q},\mathbf{v})} u^\Delta. \tag{9.2}$$

This equation shows also that the cotangent vector $\mathbf{p}$, which corresponds to the tangent vector $\mathbf{v}$ according to the transformation (9.1), is given by

$$p_\Delta = p_\Delta(q^\Gamma, v^\Gamma) = \frac{\partial}{\partial v^\Delta} L(q^\Gamma, v^\Gamma), \qquad \Delta = 1, \ldots, n. \tag{9.3}$$

From (9.3) the change of variables from (q^Γ, v^Γ) to (q^Γ, p_Γ) is a local diffeomorphism if and only if the Hessian matrix $[\partial^2 L/\partial v^\Delta \partial v^\Gamma]$ is nonsingular. When this regularity condition holds, we can invert the system (9.3) and express v^Δ as a function of q^Γ and p_Γ:

$$v^\Delta = v^\Delta(q^\Gamma, p_\Gamma), \qquad \Delta = 1, \ldots, n. \tag{9.4}$$

In particular, a curve $\{\mathbf{q} = \mathbf{q}(t), \mathbf{v} = \mathbf{v}(t)\}$ in $\mathscr{T}(\mathscr{M})$ corresponds uniquely to a curve $\{\mathbf{q} = \mathbf{q}(t), \mathbf{p} = \mathbf{p}(t)\}$ in $\mathscr{T}^*(\mathscr{M})$, and vice versa. Based on this invertible transformation, the flow problem associated with the system (7.23) on $\mathscr{T}(\mathscr{M})$ corresponds uniquely to the flow problem of a particular system on $\mathscr{T}^*(\mathscr{M})$. As we have mentioned in Section 7, the system on $\mathscr{T}^*(\mathscr{M})$ is called the *Hamiltonian system.* Before deriving that system, we consider first the physical meaning of the transformation (9.3) in more detail.

Recall that in the case of time-independent constraint, the Lagrangian function $L(q^\Delta, v^\Delta)$ is just the difference of the kinetic energy $E = \frac{1}{2}g_{\Gamma\Delta}(q)v^\Delta v^\Gamma$ and the potential energy $V = V(q^\Delta)$, viz.,

$$L(q^\Delta, v^\Delta) = \tfrac{1}{2}g_{\Delta\Gamma}(q^\Delta)v^\Delta v^\Gamma - V(q^\Delta). \tag{9.5}$$

In this case (9.3) reduces to

$$p_\Delta = p_\Delta(q^\Delta, v^\Delta) = g_{\Delta\Gamma}(q^\Delta)v^\Gamma, \qquad \Delta = 1, \ldots, n. \tag{9.6}$$

The meaning of (9.6) is clear both mathematically and physically. In the mathematical interpretation (9.6) means that $\mathbf{p}$ is the covariant vector corresponding to the contravariant vector $\mathbf{v}$ relative to the inertia metric $\mathbf{g}$. In the physical interpretation (9.6) means that $\mathbf{p}$ plays the role of the momentum corresponding to the generalized velocity $\mathbf{v}$. By virtue of this interpretation the image of $\dot{\mathbf{q}}$ defined by (9.3) is called the *generalized momentum* of the constrained system.

Now using the transformation (9.3) and its inverse (9.4), we can regard any function of $(\mathbf{q}, \mathbf{v})$ as a function of $(\mathbf{q}, \mathbf{p})$, and vice versa. With this transformation in mind we define the *Hamiltonian function* $H = H(\mathbf{q}, \mathbf{p})$ by

$$H = \mathbf{p}(\mathbf{v}) - L = p_\Delta v^\Delta - L. \tag{9.7}$$

In the simple case when L is given by (9.5), the Hamiltonian function is simply the total energy; i.e.,

$$H = \tfrac{1}{2}g_{\Delta\Gamma}v^\Delta v^\Gamma + V. \tag{9.8}$$

But it is important to note that, in both (9.7) and (9.8) the function H is regarded as a function of $\mathbf{q}$ and $\mathbf{p}$. Specifically, (9.7) means that

$$H(q^\Gamma, p_\Delta) = p_\Delta v^\Delta(q^\Delta, p_\Delta) - L\big(q^\Delta, v^\Gamma(q^\Delta, p_\Delta)\big), \tag{9.9}$$

where $v^\Delta(q^\Gamma, p_\Gamma)$ is given by the inverse transformation (9.4). Using the Hamiltonian function, we claim that the system on $\mathscr{T}^*(\mathscr{M})$ corresponding to the system (7.23) is given by

$$\frac{dq^\Delta}{dt} = \frac{\partial H}{\partial p_\Delta}, \qquad \frac{dp_\Delta}{dt} = -\frac{\partial H}{\partial q^\Delta}. \tag{9.10}$$

This is the *Hamiltonian system* on the phase space $\mathscr{T}^*(\mathscr{M})$.

To prove that (9.10) is equivalent to (7.23), we have to show that a curve $\big(\mathbf{q}(t), \mathbf{v}(t)\big)$ satisfies (7.23) if and only if its Legendre transformation $\big(\mathbf{q}(t), \mathbf{p}(t)\big)$ satisfies (9.10). We establish necessity first. From (9.9) the partial derivatives of H and L are related by

$$\frac{\partial H}{\partial q^{\Delta}} = p_{\Gamma}\frac{\partial v^{\Gamma}}{\partial q^{\Delta}} - \frac{\partial L}{\partial q^{\Delta}} - \frac{\partial L}{\partial v^{\Gamma}}\frac{\partial v^{\Gamma}}{\partial q^{\Delta}} = -\frac{\partial L}{\partial q^{\Delta}},$$
$$\frac{\partial H}{\partial p_{\Delta}} = v^{\Delta} + p_{\Gamma}\frac{\partial v^{\Gamma}}{\partial p_{\Delta}} - \frac{\partial L}{\partial v^{\Gamma}}\frac{\partial v^{\Gamma}}{\partial p_{\Delta}} = v^{\Delta}, \tag{9.11}$$

where we have used the condition (9.3). From (9.11) and (7.23) we obtain (9.10) directly. Thus necessity is proved. The proof of sufficiency is similar. From (9.11) and (9.10) we obtain

$$\frac{dq^{\Delta}}{dt} = v^{\Delta}, \qquad \frac{d}{dt}\left(\frac{\partial L}{\partial v^{\Delta}}\right) - \frac{\partial L}{\partial q^{\Delta}} = 0. \tag{9.12}$$

Thus (7.23) follows.

Unlike the system (7.23) on $\mathscr{T}(\mathscr{M})$, the Hamiltonian system (9.10) is rather symmetrical with respect to the dependent variables q^{Δ} and p_{Δ}, and for this reason its mathematical structure is easier to analyze. As before we regard (9.10) as a flow problem generated by the vector field $\mathbf{h}$ on $\mathscr{T}^*(\mathscr{M})$ with component form

$$\mathbf{h} = \frac{\partial H}{\partial p_{\Delta}}\mathbf{H}_{\Delta} - \frac{\partial H}{\partial q^{\Delta}}\mathbf{K}^{\Delta} \tag{9.13}$$

relative to the coordinate system (q^{Δ}, p_{Δ}). Since the components of $\mathbf{h}$ are determined by a single function H, they are not entirely arbitrary. We can explain the special nature of the vector field $\mathbf{h}$ by using the symplectic structure on $\mathscr{T}^*(\mathscr{M})$.

We recall that the symplectic form $\boldsymbol{\omega}$ is a nonsingular exact 2-form on $\mathscr{T}^*(\mathscr{M})$. At each point $(\mathbf{q}, \mathbf{p}) \in \mathscr{T}^*(\mathscr{M})$, $\boldsymbol{\omega}(\mathbf{q}, \mathbf{p})$ gives rise to an isomorphism of the tangent space $\mathscr{T}^*(\mathscr{M})_{(\mathbf{q},\mathbf{p})}$ with the cotangent space $\mathscr{T}^*(\mathscr{M})^*_{(\mathbf{q},\mathbf{p})}$. Hence we can apply $\boldsymbol{\omega}$ to the vector field $\mathbf{h}$ and obtain a 1-form on $\mathscr{T}^*(\mathscr{M})$. This 1-form turns out to be just the gradient dH of the Hamiltonian function, and thus is an exact 1-form. Indeed, the image of $\mathbf{h}$ under $\boldsymbol{\omega}$ can be obtained easily by using (8.23):

$$\begin{aligned}\boldsymbol{\omega}(\mathbf{h}) &= \frac{\partial H}{\partial p_{\Delta}}\boldsymbol{\omega}(\mathbf{H}_{\Delta}) - \frac{\partial H}{\partial q^{\Delta}}\boldsymbol{\omega}(\mathbf{K}^{\Delta})\\ &= \frac{\partial H}{\partial p_{\Delta}}\mathbf{K}_{\Delta} + \frac{\partial H}{\partial q^{\Delta}}\mathbf{H}^{\Delta} = dH.\end{aligned} \tag{9.14}$$

This is the key property of the Hamiltonian system (9.10).

In general we call a vector field on $\mathscr{T}^*(\mathscr{M})$ a *Hamiltonian vector field* if its image under $\boldsymbol{\omega}$ is an exact 1-form. From the Poincaré lemma (cf. Section 52, IVT-2) a closed form is locally exact, and vice versa, so we call a vector field a *locally Hamiltonian vector field* if its image under $\boldsymbol{\omega}$ is a closed 1-form. The Hamiltonian system (9.10) is, then, the flow problem associated with a particular Hamiltonian vector field.

We can characterize a locally Hamiltonian vector field $\mathbf{k}$ in general by the following:

Theorem. A necessary and sufficient condition for a vector field $\mathbf{k}$ on $\mathscr{T}^*(\mathscr{M})$ to be a locally Hamiltonian vector field is that the Lie derivative of the symplectic form $\boldsymbol{\omega}$ with respect to $\mathbf{k}$ vanishes,

$$\underset{\mathbf{k}}{\mathscr{L}}\,\boldsymbol{\omega} = \mathbf{0}, \tag{9.15}$$

or, equivalently, the flow of $\mathbf{k}$ preserves the 2-form $\boldsymbol{\omega}$.

Proof. Recall that the Lie derivative and the exterior derivative commute [cf. (51.30) in Section 51, IVT-2]. Consequently, (9.15) is equivalent to the condition that the Lie derivative of the canonical 1-form $\boldsymbol{\theta}$ be closed; i.e.,

$$\underset{\mathbf{k}}{\mathscr{L}}\,\boldsymbol{\omega} = \underset{\mathbf{k}}{\mathscr{L}}\,d\boldsymbol{\theta} = d\underset{\mathbf{k}}{\mathscr{L}}\,\boldsymbol{\theta} = \mathbf{0}. \tag{9.16}$$

Now let $\mathbf{k}$ be given by the component form

$$\mathbf{k} = \alpha^\Delta \mathbf{H}_\Delta + \beta_\Delta \mathbf{K}^\Delta \tag{9.17}$$

relative to the coordinate system (q^Δ, p_Δ). Then from (49.42) in Section 49, IVT-2, and (8.11) we have

$$\begin{aligned}\underset{\mathbf{k}}{\mathscr{L}}\,\boldsymbol{\theta} &= \left(\beta_\Delta + \frac{\partial \alpha^\Gamma}{\partial q^\Delta} p_\Gamma\right)\mathbf{H}^\Delta + \frac{\partial \alpha^\Gamma}{\partial p_\Delta} p_\Gamma \mathbf{K}_\Delta = \beta_\Delta \mathbf{H}^\Delta + \alpha^\Delta \mathbf{K}_\Delta + d(\alpha^\Delta p_\Delta) \\ &= \boldsymbol{\omega}(\mathbf{k}) + d(\boldsymbol{\theta}(\mathbf{k})).\end{aligned} \tag{9.18}$$

This result implies that $\mathscr{L}_{\mathbf{k}}\,\boldsymbol{\theta}$ is closed if and only if $\boldsymbol{\omega}(\mathbf{k})$ is also. Consequently, the condition (9.15) is necessary and sufficient for $\mathbf{k}$ to be a locally Hamiltonian vector field.

From the definition of the Lie derivative [cf. (49.41) in Section 49, IVT-2], it is clear that the condition (9.15) is equivalent to the condition that the flow of $\mathbf{k}$ preserves the 2-form $\boldsymbol{\omega}$. Thus the theorem is proved. □

In general we call a diffeomorphism from a domain in $\mathscr{T}^*(\mathscr{M})$ onto another domain in $\mathscr{T}^*(\mathscr{M})$ a *canonical transformation* if it preserves the symplectic form $\boldsymbol{\omega}$. Then the flow induced by a locally Hamiltonian vector field is a 1-parameter family of canonical transformations. Since the Hamiltonian system (9.10) corresponds to the flow problem of a Hamiltonian vector field, its solution, which determines the trajectories of the dynamical system, can be viewed as a 1-parameter family of canonical transformations such that the initial values $(q^\Delta(0), p_\Delta(0))$ are mapped to the points $(q^\Delta(t), p_\Delta(t))$ of the trajectories for each t.

In Section 7 we have defined the concept of an *integral* of a dynamical system, and we have pointed out that in the case of time-independent constraint one such integral is just the Hamiltonian function H; cf. the Jacobi integral (7.19). We can prove this fact by using the Hamiltonian system (9.10) also. Indeed, on any trajectory the time derivative of H vanishes, since we have

$$\frac{dH}{dt} = \frac{\partial H}{\partial q^\Delta}\frac{dq^\Delta}{dt} + \frac{\partial H}{\partial p_\Delta}\frac{dp_\Delta}{dt} = \frac{\partial H}{\partial q^\Delta}\frac{\partial H}{\partial p_\Delta} - \frac{\partial H}{\partial p_\Delta}\frac{\partial H}{\partial q^\Delta} = 0. \tag{9.19}$$

In general an integral $K = K(q^\Delta, p_\Delta)$ must satisfy the condition

$$0 = \frac{dK}{dt} = \frac{\partial K}{\partial q^\Delta}\frac{\partial H}{\partial p_\Delta} - \frac{\partial K}{\partial p_\Delta}\frac{\partial H}{\partial q^\Delta} \equiv \{K, H\}. \tag{9.20}$$

The scalar field $\{K, H\}$ defined by this equation is called the *Poisson bracket* of K and H. It is easy to prove that the Poisson bracket $\{K, H\}$ is independent of the choice of the coordinate system (q^Δ, p_Δ), since its value can be determined by the coordinate-free formula

$$\{K, H\} = dK(\boldsymbol{\omega}^{-1}(dH)) = \omega^{-1}(dK, dH), \tag{9.21}$$

where $\boldsymbol{\omega}^{-1}$, being the inverse of the symplectic form $\boldsymbol{\omega}$, is a canonical skew-symmetric contravariant tensor field of order 2 on $\mathscr{T}^*(\mathscr{M})$.

From (9.20) or (9.21) we see that the Poisson bracket is a skew-symmetric bilinear operation. In particular, we have

$$\{H, H\} = 0, \tag{9.22}$$

which is equivalent to (9.19); i.e., H is the Jacobi integral of the dynamical system. The skew symmetry of the Poisson bracket is an important condition in the coordinate-free interpretation of the flow generated by the Hamiltonian system.

Mathematically, the determination of a general solution for the system (9.10) is equivalent to the problem of finding $(2n - 1)$ independent integrals. Indeed, from any $(2n - 1)$ independent integrals $K_1, \ldots, K_{(2n-1)}$ we can define the path of a trajectory by the algebraic equations

$$K_A(q^\Delta, p_\Delta) = \text{const}, \qquad A = 1, \ldots, (2n - 1). \tag{9.23}$$

Then the parametrization t on the path can be assigned by solving a single first-order differential equation. In this way we obtain a $2n$-parameter family of trajectories of (9.10),

$$q^\Delta = q^\Delta(t, K_1, \ldots, K_{2n}), \qquad p_\Delta = p_\Delta(t, K_1, \ldots, K_{2n}), \tag{9.24}$$

where K_{2n} is the additioned constant of integration from the solution of the first-order equation. Since (9.24) contains $2n$ arbitrary constants, $K_1, \ldots, K_{2n}$, it is a general solution of (9.10).

Conversely, if we are given a general solution of (9.10) in the form (9.24), then we can invert it and get a system of algebraic equations

$$K_A = \tilde{K}_A(t, q^\Delta, p_\Delta), \qquad A = 1, \ldots, 2n. \tag{9.25}$$

We call a function $\tilde{K}(t, q^\Delta, p_\Delta)$ a *time-dependent integral* if its total time derivative vanishes on any trajectory of (9.10), viz.,

$$\frac{d\tilde{K}}{dt} = \frac{\partial \tilde{K}}{\partial q^\Delta}\frac{\partial H}{\partial p_\Delta} - \frac{\partial \tilde{K}}{\partial p_\Delta}\frac{\partial H}{\partial q^\Delta} + \frac{\partial \tilde{K}}{\partial t} = \{\tilde{K}, H\} + \frac{\partial \tilde{K}}{\partial t} = 0. \tag{9.26}$$

In this sense the functions $\tilde{K}_A(t, q^\Delta, p_\Delta)$ in (9.25) are $2n$ time-dependent integrals. Eliminating the parameter t from the system (9.25), we obtain a system of $(2n - 1)$ (time-independent) integrals.

Now the problem of finding an integral $K(q^\Delta, p_\Delta)$ in general can be interpreted in the following way: We consider the Hamiltonian vector field $\mathbf{k} = \boldsymbol{\omega}^{-1}(dK)$ induced by a function $K(q^\Delta, p_\Delta)$. Then the flow of $\mathbf{k}$ must preserve the symplectic form $\boldsymbol{\omega}$. Suppose that the flow also preserves the Hamiltonian function H; i.e.,

$$\begin{aligned}\frac{dH}{d\tau} &= \frac{\partial H}{\partial q^\Delta}\frac{dq^\Delta}{d\tau} + \frac{\partial H}{\partial p_\Delta}\frac{dp_\Delta}{d\tau} = \frac{\partial H}{\partial q^\Delta}\frac{\partial K}{\partial p_\Delta} - \frac{\partial H}{\partial p_\Delta}\frac{\partial K}{\partial q^\Delta} \\ &= \{H, K\} = 0,\end{aligned} \tag{9.27}$$

where τ denotes the parameter of the flow of $\mathbf{k}$. Then K satisfies the condition (9.20), since the Poisson bracket is skew symmetric. Thus K is an integral. Conversely, if K is an integral, then from (9.20) we see that (9.27) holds.

Hence the flow of **k** preserves the Hamiltonian function. Therefore a function K is an integral if and only if the flow of its corresponding Hamiltonian vector field **k** preserves the Hamiltonian function H.

Summarizing the preceding analysis, we have the following important result:

Theorem. Let K be any (time-independent) integral. Then the flow generated by the locally Hamiltonian vector field $\mathbf{k} \equiv \boldsymbol{\omega}^{-1}(dK)$ is a local 1-parameter group of canonical transformations which preserves the Hamiltonian function H. Conversely, let $\boldsymbol{\rho}_\tau$ be any local 1-parameter group of canonical transformations which preserves the Hamiltonian function H. Then the generator **k** of $\boldsymbol{\rho}_\tau$ corresponds to a closed 1-form $\boldsymbol{\omega}(\mathbf{k})$ via the canonical 2-form $\boldsymbol{\omega}$, and the potential function K of $\boldsymbol{\omega}(\mathbf{k})$ [i.e., $\boldsymbol{\omega}(\mathbf{k}) = dK$] is a (time-independent) integral.

From this theorem we see that the dynamical problem of solving the Hamiltonian system (9.10) is mathematically equivalent to the geometric problem of finding the symmetry group of the symplectic form $\boldsymbol{\omega}$ and the Hamiltonian function H. Indeed, each integral corresponds to a 1-parameter group in the symmetry group of $(\boldsymbol{\omega}, H)$, and vice versa.

10. The Legendre Transformation and the Hamiltonian System II: The Time-Dependent Case

When the constraint of the dynamical system is time dependent, we define the $(2n + 1)$-dimensional manifold $\widetilde{\mathscr{T}(\mathscr{M})}$ by

$$\widetilde{\mathscr{T}(\mathscr{M})} = \bigcup_{t \in \mathscr{R}} \mathscr{T}(\mathscr{M}^t), \tag{10.1}$$

where $\mathscr{M}^t$ denotes the constraint surface at time t, and where $\mathscr{T}(\mathscr{M}^t)$ denotes the tangent bundle of $\mathscr{M}^t$. Like the manifold structure of $\widetilde{\mathscr{M}}$ defined in Section 3, the manifold structure of $\widetilde{\mathscr{T}(\mathscr{M})}$ is determined by the coordinates (q^Δ, v^Δ, t), where (q^Δ, v^Δ) are the coordinates in $\mathscr{T}(\mathscr{M}^t)$ introduced in the preceding section. Using the manifold $\widetilde{\mathscr{T}(\mathscr{M})}$ and the coordinate system (q^Δ, v^Δ, t), we can reduce Lagrange's equations (7.3) to a system of first-order differential equations:

$$\frac{dq^\Delta}{dt} = v^\Delta, \qquad \frac{\partial^2 L}{\partial v^\Delta\, \partial v^\Gamma}\frac{dv^\Gamma}{dt} = \frac{\partial L}{\partial q^\Delta} - \frac{\partial^2 L}{\partial v^\Delta\, \partial q^\Gamma} v^\Gamma - \frac{\partial^2 L}{\partial v^\Delta\, \partial t}, \tag{10.2}$$

where the Lagrangian function $L = L(q^\Delta, v^\Delta, t)$ now depends on t. A trajectory of the dynamical system is a curve in $\widetilde{\mathscr{E}(\mathscr{M})}$ with coordinate representation

$$q^\Delta = q^\Delta(t), \qquad v^\Delta = v^\Delta(t), \qquad t = t, \tag{10.3}$$

such that $q^\Delta(t)$ and $v^\Delta(t)$ satisfy (10.2).

We define the phase space $\widetilde{\mathscr{E}^*(\mathscr{M})}$ similarly by

$$\widetilde{\mathscr{E}^*(\mathscr{M})} = \bigcup_{t\in\mathscr{R}} \mathscr{E}^*(\mathscr{M}^t), \tag{10.4}$$

which is a manifold of dimension $(2n + 1)$. A coordinate system for this manifold is (q^Δ, p_Δ, t), where (q^Δ, p_Δ) are defined in $\mathscr{E}^*(\mathscr{M}^t)$ as before for each $t \in \mathscr{R}$. We define the Legendre transformation from $\widetilde{\mathscr{E}(\mathscr{M})}$ to $\widetilde{\mathscr{E}^*(\mathscr{M})}$ by the condition (9.1), except that now L may be time dependent, viz.,

$$\mathbf{p}(\mathbf{u}) = \frac{d}{ds} L(\mathbf{q}, \mathbf{v} + s\mathbf{u}, t)\,\big|_{s=0} \tag{10.5}$$

for all $\mathbf{u} \in \mathscr{M}_{\mathbf{q}}{}^t$. In terms of the coordinates (q^Δ, v^Δ, t) and (q^Δ, p_Δ, t) in $\widetilde{\mathscr{E}(\mathscr{M})}$ and $\widetilde{\mathscr{E}^*(\mathscr{M})}$, respectively, the Legendre transformation is given by

$$p_\Delta = p_\Delta(q^\Lambda, v^\Lambda, t) = \frac{\partial L(q^\Lambda, v^\Lambda, t)}{\partial v^\Delta}, \qquad \Delta = 1, \ldots, n. \tag{10.6}$$

We assume again that L is a regular Lagrangian, so that the Legendre transformation is invertible with respect to $\mathbf{v}$ and $\mathbf{p}$, and we denote the inverse by

$$v^\Delta = v^\Delta(q^\Lambda, p_\Lambda, t), \qquad \Delta = 1, \ldots, n. \tag{10.7}$$

Consequently, a curve in $\widetilde{\mathscr{E}(\mathscr{M})}$ of the form (10.3) corresponds uniquely to a curve in $\widetilde{\mathscr{E}^*(\mathscr{M})}$ of the form

$$q^\Delta = q^\Delta(t), \qquad p_\Delta = p_\Delta(t), \qquad t = t, \tag{10.8}$$

and vice versa.

We define the Hamiltonian function $H(\mathbf{q}, \mathbf{p}, t)$ on the phase space $\widetilde{\mathscr{E}^*(\mathscr{M})}$ by (9.7), except that now both $\mathbf{v}$ and L may depend explicitly on t, viz.,

$$H(q^\Lambda, p_\Lambda, t) = p_\Delta v^\Delta(q^\Lambda, p_\Lambda, t) - L\big(q^\Lambda, v^\Gamma(q^\Lambda, p_\Lambda, t), t\big). \tag{10.9}$$

We claim that the governing system for the trajectories in phase space is still the Hamiltonian system (9.10), provided that the time-dependent Hamiltonian function defined by (10.9) is used. The proof is more or less the same as before. First, if $(\mathbf{q}(t), \mathbf{v}(t), t)$ satisfies Lagrange's system (10.2), then under the Legendre transformation the image curve $(\mathbf{q}(t), \mathbf{p}(t), t)$ satisfies the Hamiltonian system (9.10), because by the chain rule the partial derivatives of L and H are still related by (9.11). Conversely, if $(\mathbf{q}(t), \mathbf{p}(t), t)$ satisfies (9.10), then from (9.11) and (9.10) we obtain (9.12) as before. Thus the preimage $(\mathbf{q}(t), \mathbf{v}(t), t)$ satisfies (10.2).

The main difference between the time-independent case and the time-dependent case is in the structure of phase space. In the time-independent case we have shown in Section 8 that the phase space $\mathscr{C}^*(\mathscr{M})$ is a symplectic manifold equipped with the canonical forms $\boldsymbol{\theta}$ and $\boldsymbol{\omega}$. Now in the time-dependent case we define a 1-form $\tilde{\boldsymbol{\theta}}$ on $\widetilde{\mathscr{C}^*(\mathscr{M})}$ by

$$\tilde{\boldsymbol{\theta}} = p_\Delta \, dq^\Delta - H\, dt = \boldsymbol{\theta}^t - H\, dt \tag{10.10}$$

and a 2-form $\tilde{\boldsymbol{\omega}}$ on $\widetilde{\mathscr{C}^*(\mathscr{M})}$ by

$$\tilde{\boldsymbol{\omega}} = d\tilde{\boldsymbol{\theta}} = dp_\Delta \wedge dq^\Delta - dH \wedge dt = \boldsymbol{\omega}^t - dH \wedge dt. \tag{10.11}$$

Notice that the tangential parts $\boldsymbol{\theta}^t$ and $\boldsymbol{\omega}^t$ of $\tilde{\boldsymbol{\theta}}$ and $\tilde{\boldsymbol{\omega}}$, respectively, are just the canonical forms of $\mathscr{C}^*(\mathscr{M}^t)$ for each $t \in \mathscr{R}$. The time parts $H\,dt$ and $dH \wedge dt$ of $\tilde{\boldsymbol{\theta}}$ and $\tilde{\boldsymbol{\omega}}$, however, are clearly not canonical, since they depend on the Hamiltonian function H as well as on the choice of the coordinate system (q^Δ, p_Δ, t) in $\widetilde{\mathscr{C}^*(\mathscr{M})}$.

From (10.4) the t coordinate curves relative to any coordinate system (q^Δ, p_Δ, t) are not intrinsic to $\widetilde{\mathscr{C}^*(\mathscr{M})}$; i.e., the phase space $\widetilde{\mathscr{C}^*(\mathscr{M})}$ is not a product space of a particular symplectic manifold with $\mathscr{R}$. The forms $\tilde{\boldsymbol{\theta}}$ and $\tilde{\boldsymbol{\omega}}$ are invariant under a change of coordinates from (q^Δ, p_Δ, t) to $(\bar{q}^\Delta, \bar{p}_\Delta, t)$ if and only if the relation between (q^Δ) and $(\bar{q}^\Delta)$ is independent of t. [When the coordinates (q^Δ) and $(\bar{q}^\Delta)$ are required to be related by a time-independent coordinate transformation only, the manifolds $\tilde{\mathscr{M}}$, $\widetilde{\mathscr{C}(\mathscr{M})}$, and $\widetilde{\mathscr{C}^*(\mathscr{M})}$ are, in effect, product spaces.] We now assume that a particular coordinate system (q^Δ, p_Δ, t) is chosen, and that other admissible coordinate systems $(\bar{q}^\Delta, \bar{p}_\Delta, t)$ may be obtained from this system by means of a time-independent coordinate transformation only. Under this assumption the forms $\tilde{\boldsymbol{\theta}}$ and $\tilde{\boldsymbol{\omega}}$ defined by (10.10) and (10.11) are invariant under any change of (admissible) coordinate system.

We now show that the 2-form $\tilde{\boldsymbol{\omega}}$ plays a role similar to that of the

2-form $\boldsymbol{\omega}$ with respect to the flow generated by the Hamiltonian system. Specifically, relative to the vector field $\tilde{\mathbf{h}}$ corresponding to the Hamiltonian system (9.10), i.e.,

$$\tilde{\mathbf{h}} = \frac{\partial H}{\partial p_\Delta}\mathbf{H}_\Delta - \frac{\partial H}{\partial q^\Delta}\mathbf{K}^\Delta + 1\mathbf{h}_{2n+1}, \tag{10.12}$$

where $\mathbf{h}_{2n+1}$ denotes the tangent vector of the t coordinate curve in $\widetilde{\mathscr{C}^*(\mathscr{M})}$, we have

$$\mathop{\mathscr{L}}_{\tilde{\mathbf{h}}} \tilde{\boldsymbol{\omega}} = \mathbf{0} \tag{10.13}$$

or, equivalently, the flow of $\tilde{\mathbf{h}}$ preserves the 2-form $\tilde{\boldsymbol{\omega}}$.

To prove (10.13), we follow the same argument as before by showing that the Lie derivative of the 1-form $\tilde{\boldsymbol{\theta}}$ is closed. Indeed, from (10.12), (10.10), and the formula (49.42) in IVT-2 we have

$$\begin{aligned}\mathop{\mathscr{L}}_{\tilde{\mathbf{h}}} \tilde{\boldsymbol{\theta}} &= \left[-\frac{\partial H}{\partial q^\Delta} + \frac{\partial}{\partial q^\Delta}\left(p_\Gamma \frac{\partial H}{\partial p_\Gamma}\right)\right] dq^\Delta + \left[-\frac{\partial H}{\partial p_\Delta} + \frac{\partial}{\partial p_\Delta}\left(p_\Gamma \frac{\partial H}{\partial p_\Gamma}\right)\right] dp_\Delta \\ &\quad + \left[-\frac{\partial H}{\partial t} + \frac{\partial}{\partial t}\left(p_\Gamma \frac{\partial H}{\partial p_\Gamma}\right)\right] dt = d[\tilde{\boldsymbol{\theta}}(\tilde{\mathbf{h}})]. \end{aligned} \tag{10.14}$$

Consequently, (10.13) follows, since the exterior derivative and the Lie derivative commute.

Since the forms $\tilde{\boldsymbol{\theta}}$ and $\tilde{\boldsymbol{\omega}}$ depend explicitly on the Hamiltonian function H, we cannot use them to characterize other Hamiltonian vector fields, say the field

$$\tilde{\mathbf{k}} = \frac{\partial K}{\partial p_\Delta}\mathbf{H}_\Delta - \frac{\partial K}{\partial q^\Delta}\mathbf{K}^\Delta + 1\mathbf{h}_{2n+1} \tag{10.15}$$

induced by a function K. We can show by direct calculation that the flow of $\tilde{\mathbf{k}}$ preserves a 2-form $\tilde{\boldsymbol{\omega}}_K$ which is defined by (10.11) with H replaced by K. However, the flow of $\tilde{\mathbf{k}}$ generally does not preserve the 2-form $\tilde{\boldsymbol{\omega}} = \tilde{\boldsymbol{\omega}}_H$.

Another new feature of the time-dependent case is the fact that the Hamiltonian function H need not be an integral of the dynamical system. Indeed, we can calculate the rate of change of H along any trajectory of (9.10) by

$$\begin{aligned}\frac{dH}{dt} &= \frac{\partial H}{\partial q^\Delta}\frac{dq^\Delta}{dt} + \frac{\partial H}{\partial p_\Delta}\frac{dp_\Delta}{dt} + \frac{\partial H}{\partial t} \\ &= \frac{\partial H}{\partial q^\Delta}\frac{\partial H}{\partial p_\Delta} - \frac{\partial H}{\partial p_\Delta}\frac{\partial H}{\partial q^\Delta} + \frac{\partial H}{\partial t} = \frac{\partial H}{\partial t}. \end{aligned} \tag{10.16}$$

An integral K, whether or not it be time dependent, is defined by the condition $dK/dt = 0$, viz.,

$$\frac{dK}{dt} = \frac{\partial K}{\partial t} + \frac{\partial K}{\partial q^\Delta}\frac{dq^\Delta}{dt} + \frac{\partial K}{\partial p_\Delta}\frac{dp_\Delta}{dt}$$
$$= \frac{\partial K}{\partial t} + \frac{\partial K}{\partial q^\Delta}\frac{\partial H}{\partial p_\Delta} - \frac{\partial K}{\partial p_\Delta}\frac{\partial H}{\partial q^\Delta} = 0. \tag{10.17}$$

Also, it is no longer true that the Hamiltonian function H is preserved by the flow of the Hamiltonian vector field $\tilde{\mathbf{k}}$ induced by an integral K. We can calculate directly from (10.15) and (10.17) that

$$\underset{\tilde{\mathbf{k}}}{\mathscr{L}} H = \frac{\partial H}{\partial t} + \frac{\partial H}{\partial q^\Delta}\frac{\partial K}{\partial p_\Delta} - \frac{\partial H}{\partial p_\Delta}\frac{\partial K}{\partial q_\Delta} = \frac{\partial H}{\partial t} + \frac{\partial K}{\partial t}, \tag{10.18}$$

where $\mathscr{L}_{\tilde{\mathbf{k}}} H$ is just the rate of change of H along any trajectory of $\tilde{\mathbf{k}}$. By the same token if K is a function of (q^Δ, p_Δ, t) in general, then we have

$$\underset{\tilde{\mathbf{k}}}{\mathscr{L}} H + \underset{\tilde{\mathbf{h}}}{\mathscr{L}} K = \frac{\partial H}{\partial t} + \frac{\partial K}{\partial t} = \underset{\tilde{\mathbf{h}}}{\mathscr{L}} H + \underset{\tilde{\mathbf{k}}}{\mathscr{L}} K. \tag{10.19}$$

The preceding analysis shows that the problem of finding integrals for the dynamical system no longer corresponds to the geometric problem of finding the transformation group which preserves the Hamiltonian function H and the 2-form $\tilde{\boldsymbol{\omega}} = \tilde{\boldsymbol{\omega}}_H$. For that reason we now introduce the new concept of *contact transformation* in order to interpret the flow problem of the Hamiltonian system for the time-dependent case.

We call a diffeomorphism $\tilde{\boldsymbol{\varphi}}: \widetilde{\mathscr{T}^*(\mathscr{M})} \to \widetilde{\mathscr{T}^*(\mathscr{M})}$, which is formed by a 1-parameter family of diffeomorphisms $\boldsymbol{\varphi}^t: \mathscr{T}^*(\mathscr{M}^t) \to \mathscr{T}^*(\mathscr{M}^t)$, $t \in \mathscr{R}$, a *contact transformation* if there is a function $F = F(q^\Delta, p_\Delta, t)$ on $\widetilde{\mathscr{T}^*(\mathscr{M})}$ such that

$$[\operatorname{grad} \tilde{\boldsymbol{\varphi}}](\tilde{\boldsymbol{\omega}}) = \tilde{\boldsymbol{\omega}} + dF \wedge dt \tag{10.20}$$

or, equivalently,

$$d\big([\operatorname{grad} \tilde{\boldsymbol{\varphi}}](\tilde{\boldsymbol{\theta}}) - \tilde{\boldsymbol{\theta}} - F\,dt\big) = 0, \tag{10.21}$$

where grad $\tilde{\boldsymbol{\varphi}}$ denotes the gradient of $\tilde{\boldsymbol{\varphi}}$. (The concept of the gradient of a mapping has been defined in Section 8 in connection with the projection map $\boldsymbol{\pi}$.) Relative to the particular coordinate system (q^Δ, p_Δ, t) on $\widetilde{\mathscr{T}^*(\mathscr{M})}$

we can express the mapping $\tilde{\boldsymbol{\varphi}}$ by giving the coordinates[2] (Q^Δ, P_Δ, t) of the image point $\tilde{\boldsymbol{\varphi}}(\mathbf{q}, \mathbf{p}, t)$ as functions of the coordinates (q^Δ, p_Δ, t) of the point $(\mathbf{q}, \mathbf{p}, t)$ in the domain, viz.,

$$Q^\Delta = Q^\Delta(q^\Delta, p_\Delta, t), \qquad P_\Delta = P_\Delta(q^\Delta, p_\Delta, t), \qquad t = t. \tag{10.22}$$

Then the condition (10.20) means that

$$dP_\Delta \wedge dQ^\Delta = dp_\Delta \wedge dq^\Delta + dF \wedge dt, \tag{10.23}$$

while the condition (10.21) means that

$$d(P_\Delta\, dQ^\Delta - p_\Delta\, dq^\Delta - F\, dt) = \mathbf{0}. \tag{10.24}$$

Note. The values of the terms $H\,dt$ and $dH \wedge dt$ under grad $\tilde{\boldsymbol{\varphi}}$ are absorbed into the terms $F\,dt$ and $dF \wedge dt$. Also, (10.24) is locally equivalent to

$$P_\Delta\, dQ^\Delta - p_\Delta\, dq^\Delta - F\, dt = dW. \tag{10.25}$$

In general a time-preserving diffeomorphism of $\widetilde{\mathscr{T}^*(\mathscr{M})}$ maps a curve of the form $\big(q^\Delta(t), p_\Delta(t), t\big)$ into a curve of the form $\big(Q^\Delta(t), P_\Delta(t), t\big)$. A necessary and sufficient condition for such a diffeomorphism to be a contact transformation is that a locally Hamiltonian vector field is transformed into a locally Hamiltonian field. Specifically, the trajectories of a system of the form

$$\frac{dq^\Delta}{dt} = \frac{\partial K}{\partial p_\Delta}, \qquad \frac{dp_\Delta}{dt} = -\frac{\partial K}{\partial q^\Delta} \tag{10.26}$$

are transformed into the trajectories of a similar system

$$\frac{dQ^\Delta}{dt} = \frac{\partial G}{\partial P_\Delta}, \qquad \frac{dP_\Delta}{dt} = -\frac{\partial G}{\partial Q^\Delta}, \tag{10.27}$$

where G is related to K by

$$G = K + F. \tag{10.28}$$

This criterion can be verified most easily by using Hamilton's principle. Indeed, the trajectories of the system (10.26) are simply the extremal curves of the line integral $\int (p_\Delta\, dq^\Delta - K\, dt)$. (The concept of line integral

[2] The coordinates Q^Δ here should not be confused with the generalized force in the preceding chapter.

is defined in Section 70, IVT-2.) As a result of (10.25), that line integral is equal to the line integral $\int (P_\Delta\, dQ^\Delta - (K + F)\, dt) + \int dW$, where the last term is a constant for curves with fixed terminal points. Consequently, extremal curves, which are characterized by the system (10.26), are transformed into extremal curves, which are characterized by the system (10.27) with G given by (10.28). It is understood, of course, that K is regarded as a function of (q^Δ, p_Δ, t) in (10.26) and as a function of (Q^Δ, P_Δ, t) in (10.28) via the transformation (10.22).

Conversely, if the transformation $\tilde{\boldsymbol{\varphi}}$ possesses the property that trajectories of (10.26) are mapped into trajectories of (10.27), we put

$$P_\Delta\, dQ^\Delta - p_\Delta\, dq^\Delta - F\, dt = \boldsymbol{\Omega}; \tag{10.29}$$

then the 1-form $\boldsymbol{\Omega}$ must be such that its line integral $\int \boldsymbol{\Omega}$ depends only on the terminal points. Hence from Stokes' theorem (cf. Section 71, IVT-2) $\boldsymbol{\Omega}$ is closed.

Now the concept of a contact transformation can be applied to the flow problem of the Hamiltonian system in the following way: We seek a contact transformation $\tilde{\boldsymbol{\varphi}}$ such that the governing equations for the image curves of the trajectories become

$$\frac{dQ^\Delta}{dt} = 0, \qquad \frac{dP_\Delta}{dt} = 0. \tag{10.30}$$

Notice that the system (10.30) is a special case of (10.27) with $G = 0$. Since the solution curves of (10.30) are simply $(Q^\Delta(0), P_\Delta(0), t)$, we say that $\tilde{\boldsymbol{\varphi}}$ transforms the Hamiltonian system (9.10) to an *equilibrium system.* The inverse of such a contact transformation is just a general solution of (9.10), viz.,

$$q^\Delta = q^\Delta(Q^\Delta, P_\Delta, t), \qquad p_\Delta = p_\Delta(Q^\Delta, P_\Delta, t), \tag{10.31}$$

where (Q^Δ, P_Δ) are $2n$ arbitrary constants.

In the following section we shall consider the procedure for constructing a contact transformation in general by means of a generating function. Then the problem of finding a particular generating function, which gives rise to a contact transformation from the Hamiltonian system to an equilibrium system, can be formulated. It turns out that this generating function is governed by a first-order partial differential equation, called the *Hamilton–Jacobi equation.* This partial differential equation possesses the property that its characteristic strips are the trajectories of the Hamiltonian system, and vice versa. We shall explain this relation in detail in the following section.

11. Contact Transformations and the Hamilton–Jacobi Equation

The concept of a contact transformation $\tilde{\boldsymbol{\varphi}}: \widetilde{\mathscr{C}^*(\mathscr{M})} \to \widetilde{\mathscr{C}^*(\mathscr{M})}$ has been defined in the preceding section. Relative to a particular coordinate system (q^Δ, p_Δ, t) in $\widetilde{\mathscr{C}^*(\mathscr{M})}$ we seek functions

$$Q^\Delta = Q^\Delta(q^\Lambda, p_\Lambda, t), \tag{11.1a}$$

$$P_\Delta = P_\Delta(q^\Lambda, p_\Lambda, t) \tag{11.1b}$$

such that, locally, the condition

$$P_\Delta\, dQ^\Delta - p_\Delta\, dq^\Delta - F\, dt = dW \tag{11.2}$$

is satisfied by a certain function W. Since $\tilde{\boldsymbol{\varphi}}$ is a time-preserving diffeomorphism, the system (11.1) is invertible on $\mathscr{M}^t$, i.e., we can solve q^Δ and p_Δ from (11.1),

$$q^\Delta = q^\Delta(Q^\Lambda, P_\Lambda, t), \tag{11.3a}$$

$$p_\Delta = p_\Delta(Q^\Lambda, P_\Lambda, t), \tag{11.3b}$$

at each $t \in \mathscr{R}$.

We consider first the special case when the differentials $\{dQ^\Delta, dq^\Delta, dt\}$ are linearly independent. In this case the first n equations, (11.1a), are invertible with respect to p_Γ, so we have the intermediate relations

$$p_\Gamma = \hat{p}_\Gamma(q^\Lambda, Q^\Lambda, t), \qquad \Gamma = 1, \ldots, n. \tag{11.4}$$

When this is the case, we can express W as a function of $(q^\Lambda, Q^\Lambda, t)$. Then

$$dW = \frac{\partial W}{\partial q^\Delta}\, dq^\Delta + \frac{\partial W}{\partial Q^\Delta}\, dQ^\Delta + \frac{\partial W}{\partial t}\, dt. \tag{11.5}$$

Substituting (11.5) into (11.2), we get

$$\left(P_\Delta - \frac{\partial W}{\partial Q^\Delta}\right) dQ^\Delta - \left(p_\Delta + \frac{\partial W}{\partial q^\Delta}\right) dq^\Delta - \left(F + \frac{\partial W}{\partial t}\right) dt = 0. \tag{11.6}$$

Since the differentials $\{dQ^\Delta, dq^\Delta, dt\}$ are assumed to be linearly independent, this equation implies

$$P_\Delta = \frac{\partial W}{\partial Q^\Delta}, \tag{11.7a}$$

$$p_\Delta = -\frac{\partial W}{\partial q^\Delta}, \tag{11.7b}$$

$$F = -\frac{\partial W}{\partial t}. \tag{11.7c}$$

The assumption (11.4) requires that (11.7b) be invertible. Hence the Hessian matrix $[\partial^2 W/\partial q^\Delta \partial Q^\Gamma]$ must be nonsingular. Inverting (11.7b), we obtain a relation of the form (11.1a), which can then be substituted into (11.7a), resulting in a relation of the form (11.1b). Thus we obtain a contact transformation which transforms the Hamiltonian system (9.10) into the system

$$\frac{dQ^\Delta}{dt} = \frac{\partial K}{\partial P_\Delta}, \qquad \frac{dP_\Delta}{dt} = -\frac{\partial K}{\partial Q^\Delta}, \tag{11.8}$$

where K is given by

$$K = H + F = H - \frac{\partial W}{\partial t} \tag{11.9}$$

and is regarded as a function of (Q^Δ, P_Δ, t) via the transformation (11.1).

A contact transformation in general need not satisfy the condition of invertibility (11.4), of course. If the rank of $[\partial Q^\Delta/\partial p_\Gamma]$ is only $(n - r)$, where r is an integer $0 < r \leq n$, then we use the method of the Lagrange multiplier to generate the corresponding contact transformation. [The value $r = 0$ corresponds to the special case which satisfies the condition of invertibility (11.4).]

Rearranging the order of Q^Δ and p_Γ if necessary, we may assume that

$$\det\left[\frac{\partial(Q^1, \ldots, Q^{n-r})}{\partial(p_1, \ldots, p_{n-r})}\right] \neq 0. \tag{11.10}$$

Then the first $(n - r)$ equations in (11.1a) may be inverted, and we have the intermediate relations

$$p_a = \hat{p}_a(q^\Gamma, Q^1, \ldots, Q^{n-r}, p_{n-r+1}, \ldots, p_n, t), \qquad a = 1, \ldots, n - r. \tag{11.11}$$

Substituting these into the remaining r equations in (11.1a), we get

$$Q^b = \hat{Q}^b(q^\Gamma, Q^1, \ldots, Q^{n-r}, p_{n-r+1}, \ldots, p_n, t), \qquad b = n - r + 1, \ldots, n. \tag{11.12}$$

This system must now be entirely independent of p_b, i.e.,

$$\left[\frac{\partial(\hat{Q}^{n-r+1}, \ldots, \hat{Q}^n)}{\partial(p_{n-r+1}, \ldots, p_n)}\right] = 0; \tag{11.13}$$

otherwise, we can solve some of the remaining p_b's in terms of (q^Γ, Q^Γ, t) and thus contradict the assumption that the rank of $[\partial Q^\Delta/\partial p_\Gamma]$ is precisely $n - r$.

From (11.13) we see that there are r identities among the variables (q^Γ, Q^Γ, t), say,

$$\Omega_c(q^\Gamma, Q^\Gamma, t) = 0, \qquad c = 1, \ldots, r. \tag{11.14}$$

As before we choose a generating function $W(q^\Gamma, Q^\Gamma, t)$ and obtain (11.6). However, since (q^Γ, Q^Γ, t) are related by the identities (11.14), the differentials $\{dq^\Gamma, dQ^\Gamma, dt\}$ are no longer linearly independent. In fact we have the linear relations

$$\frac{\partial \Omega_c}{\partial q^\Gamma} dq^\Gamma + \frac{\partial \Omega_c}{\partial Q^\Gamma} dQ^\Gamma + \frac{\partial \Omega_c}{\partial t} dt = 0, \qquad c = 1, \ldots, r, \tag{11.15}$$

so that the coefficients in (11.6) are arbitrary linear combinations of the coefficients in (11.15). This result is just the expression of the Lagrange multipliers; i.e.,

$$P_\Delta - \frac{\partial W}{\partial Q^\Delta} = \lambda_1 \frac{\partial \Omega_1}{\partial q^\Delta} + \cdots + \lambda_r \frac{\partial \Omega_r}{\partial q^\Delta} = \lambda_c \frac{\partial \Omega_c}{\partial q^\Delta}, \tag{11.16a}$$

$$-\left(p_\Delta + \frac{\partial W}{\partial q^\Delta}\right) = \lambda_1 \frac{\partial \Omega_1}{\partial Q^\Delta} + \cdots + \lambda_r \frac{\partial \Omega_r}{\partial Q^\Delta} = \lambda_c \frac{\partial \Omega_c}{\partial Q^\Delta}, \tag{11.16b}$$

$$-\left(F + \frac{\partial W}{\partial t}\right) = \lambda_1 \frac{\partial \Omega_1}{\partial t} + \cdots + \lambda_r \frac{\partial \Omega_r}{\partial t} = \lambda_c \frac{\partial \Omega_c}{\partial t}, \tag{11.16c}$$

where c is summed from 1 to r. The system (11.16a), (11.16b) and the identities (11.14) jointly define the transformation (11.1) and the Lagrange multipliers λ_c, $c = 1, \ldots, r$, as functions of (q^Γ, p_Γ, t) or as functions of (Q^Γ, P_Γ, t). Under this contact transformation the Hamiltonian system (9.10) is transformed into the system (11.8), where K is now given by

$$K = H + F = H - \frac{\partial W}{\partial t} - \lambda_c \frac{\partial \Omega_c}{\partial t}, \tag{11.17}$$

which is regarded as a function of (Q^Δ, P_Δ, t).

From the preceding analysis we see that a contact transformation subject to the identities (11.14) is rather difficult to construct. Therefore we shall now return to the special case which satisfies the condition of invertibility (11.4); i.e., we assume that there are no identities of the form (11.14) among the variables (q^Δ, Q^Δ, t). Under this assumption the contact transformation generated by W transforms the Hamiltonian system (9.10) to an equilibrium system if W satisfies the partial differential equation

$$\frac{\partial W}{\partial t} - H\left(q^{\Delta}, -\frac{\partial W}{\partial q^{\Delta}}, t\right) = 0. \tag{11.18}$$

It is customary to write this equation in the form

$$\frac{\partial U}{\partial t} + H\left(q^{\Delta}, \frac{\partial U}{\partial q^{\Delta}}, t\right) = 0, \tag{11.19}$$

which is called the *Hamilton–Jacobi equation.* We seek an n-parameter family of solutions of (11.19) of the form

$$U = U(q^{\Delta}, Q^{\Delta}, t), \tag{11.20}$$

where Q^{Δ} are independent parameters in the sense that the Hessian matrix $[\partial^2 U/\partial q^{\Delta}\partial Q^{\Gamma}]$ is nonsingular. Then $W = -U$ is the generating function for the contact transformation.

Clearly, the Hamilton–Jacobi equation is a first-order partial differential equation for the unknown function

$$U = U(q^{\Delta}, t) \tag{11.21}$$

on the constraint surface $\tilde{\mathscr{M}} = \bigcup_{t \in \mathscr{R}} \mathscr{M}^t$ with coordinates (q^{Γ}, t). The solutions of such a first-order partial differential equation may be described by the *method of characteristics*, which is summarized in the appendix to this chapter.

12. The Hamilton–Jacobi Theory

Using the notations in the theory of first-order partial differential equations (cf. the appendix to this chapter), we write the Hamilton–Jacobi equation (11.19) in the form

$$\Phi(q^{\Delta}, t, U, p_{\Delta}, p) \equiv p + H(q^{\Delta}, p_{\Delta}, t) = 0, \tag{12.1}$$

where we have denoted the partial derivatives $\partial U/\partial q^{\Delta}$ and $\partial U/\partial t$ by p_{Δ} and p, respectively. For simplicity we denote a partial derivative of a function in general by a subscript, e.g., $\partial \Phi/\partial q^{\Delta} = \Phi_{q^{\Delta}}$, $\partial H/\partial t = H_t$, etc.

From (12.1) we then have

$$\Phi_{q^\Delta} = H_{q^\Delta}, \tag{12.2a}$$

$$\Phi_t = H_t, \tag{12.2b}$$

$$\Phi_U = 0, \tag{12.2c}$$

$$\Phi_{p_\Delta} = H_{p_\Delta}, \tag{12.2d}$$

$$\Phi_p = 1, \tag{12.2e}$$

where H is the Hamiltonian function.

For the Hamilton–Jacobi equation (12.1) the conditions for a Monge strip $\big(q^\Delta(\tau), t(\tau), U(\tau), p_\Delta(\tau), p(\tau)\big)$ can be read off from (A.36), viz.,

$$(q^\Delta)' = H_{p_\Delta}, \tag{12.3a}$$

$$t' = 1, \tag{12.3b}$$

$$U' = p_\Delta H_{p_\Delta} + p, \tag{12.3c}$$

$$p + H(q^\Delta, p_\Delta, t) = 0. \tag{12.3d}$$

Similarly, the conditions for a characteristic strip can be read off from (A.38), viz.,

$$(q^\Delta)' = H_{p_\Delta}, \tag{12.4a}$$

$$t' = 1, \tag{12.4b}$$

$$U' = p_\Delta H_{p_\Delta} + p, \tag{12.4c}$$

$$(p_\Delta)' = -H_{q^\Delta}, \tag{12.4d}$$

$$p' = H_t, \tag{12.4e}$$

together with the algebraic condition (12.3d), which is automatically satisfied on the whole strip if it is satisfied at an initial point. The systems (12.3) and (12.4) can be interpreted in the following way:

First, a Monge strip $\big(q^\Delta(\tau), t(\tau), U(\tau), p_\Delta(\tau), p(\tau)\big)$ corresponds to a curve $\big(q^\Delta(t), p_\Delta(t), t\big)$ in the phase space $\widetilde{\mathscr{C}^*(\mathscr{M})}$ such that

$$\begin{aligned} \tau &= t, \\ q^\Delta(\tau) &= q^\Delta(t), \\ p_\Delta(\tau) &= p_\Delta(t), \\ p(\tau) &= -H\big(q^\Delta(t), p_\Delta(t), t\big), \\ U(\tau) &= U(t), \end{aligned} \tag{12.5}$$

where $p_\Delta(t)$ are entirely arbitrary, $q^\Delta(t)$ are determined by solving the system of first-order differential equations

$$\frac{dq^\Delta}{dt} = H_{p_\Delta}(q^\Delta, p_\Delta(t), t), \qquad \Delta = 1, \ldots, n, \tag{12.6}$$

and $U(t)$ is determined by solving the single first-order differential equation

$$\frac{dU}{dt} = p_\Delta(t)H_{p_\Delta}(q^\Delta(t), p_\Delta(t), t) - H(q^\Delta(t), p_\Delta(t), t). \tag{12.7}$$

The collection of curves $(q^\Delta(t), p_\Delta(t), t)$ which corresponds to the collection of Monge strips clearly contains the trajectories of the dynamical system, since the system (12.6) is just the first half of the Hamiltonian system (9.10). However, since $p_\Delta(t)$ are entirely arbitrary, there are many curves in this collection which do not satisfy (9.10).

The other half of the Hamiltonian system is contained in (12.4). As a result, a characteristic strip corresponds precisely to a trajectory in $\widetilde{\mathscr{C}^*(\mathscr{M})}$, and vice versa. The transformation from $(q^\Delta(\tau), t(\tau), U(\tau), p_\Delta(\tau), p(\tau))$ to $(q^\Delta(t), p_\Delta(t), t)$ is again given by (12.5), except that now $(q^\Delta(t), p_\Delta(t))$ are determined by the Hamiltonian system (9.10), and then $U(t)$ is determined by the differential equation (12.7). Notice that $p(t)$ is just the negative value of the Hamiltonian function on the trajectory as shown by (12.5d). Then (12.4e) is just the rate equation (10.16), which we have observed before.

The one-to-one correspondence between a characteristic strip of the Hamilton–Jacobi equation and a trajectory of the Hamiltonian system is the central result of the Hamilton–Jacobi theory. In the appendix to this chapter we have explained the construction of a particular solution surface by a family of characteristic strips. In dynamics, however, we are more interested in the trajectories. In fact the very reason for considering the Hamilton–Jacobi equation is to find a generating function W which gives rise to a contact transformation from the Hamiltonian system to an equilibrium system. Such a generating function corresponds to an n-parameter family of solutions $U = U(q^\Delta, Q^\Delta, t)$ of the Hamilton–Jacobi equation such that the Hessian matrix $[\partial^2 U/\partial q^\Delta \partial Q^\Gamma]$ is nonsingular.

In the theory of first-order partial differential equations a concept closely related to the generating function is called a *complete integral.* For equation (A.34) in the appendix to this chapter a complete integral is an m-parameter family of solutions

$$u = u(x^1, \ldots, x^m, \lambda^1, \ldots, \lambda^m) \tag{12.8}$$

such that the Hessian matrix $[\partial^2 u/\partial x^i \partial \lambda^j]$ is nonsingular. From such a complete integral there is a standard procedure, which we shall not discuss here, for generating a much bigger family of solutions. In fact in some cases all solutions of the partial differential equation may be generated. For the Hamilton–Jacobi equation (12.1) we need only to find an n-parameter family of solutions

$$U = U(q^\Delta, Q^\Delta, t) \tag{12.9}$$

such that $[\partial^2 U/\partial q^\Delta \partial Q^\Gamma]$ is nonsingular, since we can obtain a complete integral from (12.9) by simply adding an extra parameter Q^{n+1} to U; i.e.,

$$\tilde{U}(q^\Delta, Q^\Delta, t, Q^{n+1}) = U(q^\Delta, Q^\Delta, t) + Q^{n+1}. \tag{12.10}$$

This result is a consequence of the fact that the dependent variable U does not appear explicitly in the argument of the function Φ for the Hamilton–Jacobi equation; cf. (12.2c).

Using the generating function $U(q^\Delta, Q^\Delta, t)$, we can define a contact transformation by first inverting the relation

$$p_\Delta = \frac{\partial U}{\partial q^\Delta} \tag{12.11}$$

to get

$$Q^\Delta = Q^\Delta(q^\Gamma, p_\Gamma, t). \tag{12.12}$$

Then we substitute (12.12) into the argument of $\partial U/\partial Q^\Delta$ to get

$$P_\Delta = -\frac{\partial U}{\partial Q^\Delta} = P_\Delta(q^\Gamma, p_\Gamma, t). \tag{12.13}$$

By virtue of the Hamilton–Jacobi equation the governing system for (Q^Δ, P_Δ) is an equilibrium system, which has the trivial solution

$$Q^\Delta = Q^\Delta(q^\Gamma, p_\Gamma, t) = \text{const}, \qquad P_\Delta = P_\Delta(q^\Gamma, p_\Gamma, t) = \text{const}. \tag{12.14}$$

Then the general solution of the original Hamiltonian system is given implicitly by (12.14) or, equivalently, by the inverse of the contact transformation, viz.,

$$q^\Delta = q^\Delta(Q^\Gamma, P_\Gamma, t), \qquad p_\Delta = p_\Delta(Q^\Gamma, P_\Gamma, t), \tag{12.15}$$

where (Q^Γ, P_Γ) are $2n$ independent constants.

The Hamilton–Jacobi theory establishes a one-to-one correspondence between the trajectories of the Hamiltonian system (9.10) and the characteristic strips of the Hamilton–Jacobi equation (12.1). But we recall that the trajectories of the Hamiltonian system are also in one-to-one correspondence with the trajectories of Lagrange's equations via the Legendre transformation. Finally, from Hamilton's principle, the trajectories of Lagrange's equations are the extremal curves of the variational integral (7.10) over any pair of fixed terminal points. Combining these three basic one-to-one relations, we can obtain an interpretation of the solution of the initial value problem for the Hamilton–Jacobi equation using concepts from a variational principle. This interpretation corresponds to *Huygens' principle* in optics. We shall discuss this interesting interpretation in the following section.

13. Huygens' Principle for the Hamilton–Jacobi Equation

In Section 7 we showed that an extremal curve for the integral (7.10) taken from a class of curves having a common pair of terminal points is characterized by Lagrange's equations (7.3). The extremal value of the integral then gives rise to a function of the pair of terminal points. In general, however, this function may fail to be single valued, since there may be more than one extremal curve joining a fixed pair of extremal points. In such a case the results of this section may be applied to any one sheet of the multiple-valued function.

We denote the single-valued function or a certain sheet of the multiple-valued function of the pair of terminal points by

$$G(q_0{}^\Delta, t_0, q_1{}^\Delta, t_1) = \int_{t_0}^{t_1} L\big(q^\Delta(t), \dot{q}^\Delta(t), t\big)\, dt, \tag{13.1}$$

where $\big(q^\Delta(t), t\big)$ denotes the particular extremal curve chosen to join the pair of terminal points $(q_0{}^\Delta, t_0)$ and $(q_1{}^\Delta, t_1)$. We call this function the *geodetic distance* from $(q_0{}^\Delta, t_0)$ to $(q_1{}^\Delta, t_1)$. It should be noted, however, that G is not a distance function in the usual sense (cf. Section 12, IVT-1). Indeed, G need not be positive, and it is skew symmetric rather than symmetric with respect to the pair of points $(q_0{}^\Delta, t_0)$ and $(q_1{}^\Delta, t_1)$, viz.,

$$G(q_0{}^\Delta, t_0, q_1{}^\Delta, t_1) = -G(q_1{}^\Delta, t_1, q_0{}^\Delta, t_0), \tag{13.2}$$

provided that the same extremal curve is used in calculating the two sides of (13.2).

We establish first an important property of the function G. Let

$(q_0{}^\Delta(\tau), t_0(\tau))$ and $(q_1{}^\Delta(\tau), t_1(\tau))$ be an arbitrary pair of curves with parameter τ. Then the restriction of G on the curves is a composite function of τ. We claim that the derivative of this composite function is given by

$$\frac{dG}{d\tau} = \left(p_\Delta \frac{dq^\Delta}{d\tau} - H\frac{dt}{d\tau}\right)\Big|_{t_0(\tau)}^{t_1(\tau)}, \tag{13.3}$$

where for each fixed τ, $p_\Delta(t, \tau)$ denotes the generalized momentum on the extremal curve $(q^\Delta(t, \tau), t)$ joining $(q_0{}^\Delta(\tau), t_0(\tau))$ and $(q_1{}^\Delta(\tau), t_1(\tau))$, and where $H(q^\Delta(t, \tau), p_\Delta(t, \tau), t)$ denotes similarly the Hamiltonian function on the extremal curve $(q^\Delta(t, \tau), t)$. The derivatives $dq^\Delta/d\tau$ and $dt/d\tau$, of course, are taken on the curves $(q_0{}^\Delta(\tau), t_0(\tau))$ and $(q_1{}^\Delta(\tau), t_1(\tau))$.

To prove (13.3), we calculate the composite function G directly from (13.1) and (10.9):

$$\begin{aligned} &G(q_0{}^\Delta(\tau), t_0(\tau), q_1{}^\Delta(\tau), t_1(\tau)) \\ &\quad = \int_{t_0(\tau)}^{t_1(\tau)} [\dot{q}^\Delta(t,\tau)p_\Delta(t,\tau) - H(q^\Delta(t,\tau), p_\Delta(t,\tau), t)]\, dt. \end{aligned} \tag{13.4}$$

Differentiating this equation with respect to τ, we get

$$\begin{aligned} \frac{dG}{d\tau} &= \left[(\dot{q}^\Delta p_\Delta - H)\frac{dt}{d\tau}\right]_{t_0(\tau)}^{t_1(\tau)} + \int_{t_0(\tau)}^{t_1(\tau)} \frac{\partial}{\partial\tau}[\dot{q}^\Delta p_\Delta - H]\, dt \\ &= \left[(\dot{q}^\Delta p_\Delta - H)\frac{dt}{d\tau}\right]_{t_0(\tau)}^{t_1(\tau)} \\ &\quad + \int_{t_0(\tau)}^{t_1(\tau)} \left[\dot{q}^\Delta \frac{\partial p_\Delta}{\partial\tau} + p_\Delta \frac{\partial \dot{q}^\Delta}{\partial\tau} - \frac{\partial H}{\partial q^\Delta}\frac{\partial q^\Delta}{\partial\tau} - \frac{\partial H}{\partial p_\Delta}\frac{\partial p_\Delta}{\partial\tau}\right] dt. \end{aligned} \tag{13.5}$$

Now using the fact that for each fixed τ the curve $(q^\Delta(t, \tau), p_\Delta(t, \tau), t)$ satisfies the Hamiltonian system (9.10), we can simplify the last integral in (13.5):

$$\int_{t_0(\tau)}^{t_1(\tau)} \left[p_\Delta \frac{\partial \dot{q}^\Delta}{\partial\tau} + \dot{p}_\Delta \frac{\partial q^\Delta}{\partial\tau}\right] dt = \int_{t_0(\tau)}^{t_1(\tau)} \frac{\partial}{\partial t}\left[\frac{\partial q^\Delta}{\partial\tau} p_\Delta\right] dt = \left[\frac{\partial q^\Delta}{\partial\tau} p_\Delta\right]_{t_0(\tau)}^{t_1(\tau)}. \tag{13.6}$$

Combining (13.5) and (13.6), we then get

$$\frac{dG}{dt} = \left[p_\Delta\left(\dot{q}^\Delta \frac{dt}{d\tau} + \frac{\partial q}{\partial\tau}\right) - H\frac{dt}{d\tau}\right]_{t_0(\tau)}^{t_1(\tau)} = \left[p_\Delta \frac{dq^\Delta}{d\tau} - H\frac{dt}{d\tau}\right]_{t_0(\tau)}^{t_1(\tau)}. \tag{13.7}$$

Thus (13.3) is proved.

Since the curves $\big(q_0{}^\Delta(\tau), t_0(\tau)\big)$ and $\big(q_1{}^\Delta(\tau), t_1(\tau)\big)$ are arbitrary, by using the coordinate curves of $(q_0{}^\Delta, t_0)$ and $(q_1{}^\Delta, t_1)$, we obtain

$$\frac{\partial G}{\partial q_0{}^\Delta} = -p_\Delta{}^0, \qquad \frac{\partial G}{\partial q_1{}^\Delta} = +p_\Delta{}^1, \qquad \frac{\partial G}{\partial t_0} = +H_0, \qquad \frac{\partial G}{\partial t_1} = -H_1, \tag{13.8}$$

where $p_\Delta{}^0$, H_0 and $p_\Delta{}^1$, H_1 denote the generalized momenta and the Hamiltonian functions at the terminal points $(q_0{}^\Delta, t_0)$ and $(q_1{}^\Delta, t_1)$ of the extremal curve.

From (13.8) we see that if the initial point $(q_0{}^\Delta, t_0)$ is held fixed, and if the terminal point $(q_1{}^\Delta, t_1)$ is regarded as the independent variable, then the function

$$V(q^\Delta, t) \equiv G(q_0{}^\Delta, t_0, q^\Delta, t) \tag{13.9}$$

is a solution of the Hamilton–Jacobi equation. The surface $\big(q^\Delta, t, V(q^\Delta, t)\big)$ of this particular solution is called the *integral conoid* with vertex at $(q_0{}^\Delta, t_0)$. This conoid is formed by the characteristic strips whose initial points correspond to the tangent planes of the Monge cone at $(q_0{}^\Delta, t_0)$. Consequently, this solution has a singularity at the initial point.

Next, we develop a procedure for generating a regular solution surface which has no singularity. To remove the fixed initial point, we now consider a fixed initial manifold of dimension n. We use an algebraic equation

$$U(q_0{}^\Delta, t_0) = 0 \tag{13.10}$$

to identify this manifold.

Relative to this initial manifold an extremal curve $(q^\Delta(t), t)$ joining $(q_0{}^\Delta, t_0)$ to $(q_1{}^\Delta, t_1)$ is called a *transversal extremal curve* if it satisfies the following condition: We keep the terminal point $(q_1{}^\Delta, t_1)$ fixed, but we allow the initial point to vary on an arbitrary curve $\big(q_0{}^\Delta(\tau), t_0(\tau)\big)$ in the initial manifold. Then we require that the geodetic distance $G\big(q_0{}^\Delta(\tau), t_0(\tau), q_1{}^\Delta, t_1\big)$ have an extremal value at the transversal extremal curve. That is, we require that the transversal extremal curve satisfy

$$\left.\frac{dG(q_0{}^\Delta(\tau), t_0(\tau), q_1{}^\Delta, t_1)}{d\tau}\right|_{(q_0{}^\Delta, t_0)} = 0 \tag{13.11}$$

relative to all curves belonging to the initial manifold and passing through the particular point $(q_0{}^\Delta, t_0)$.

Using (13.8), we can express the condition (13.11) by

$$-p_\Delta{}^0 \frac{dq_0{}^\Delta}{d\tau} + H_0 \frac{dt_0}{d\tau} = 0, \tag{13.12}$$

which means that the covector $-p_\Delta{}^0\mathbf{h}^\Delta + H_0\mathbf{h}^{n+1}$ is orthogonal to an arbitrary tangent vector of the initial manifold at $(q_0{}^\Delta, t_0)$. Since the normal covector of the initial manifold (13.10) is given, in general, by $(\partial U/\partial q_0{}^\Delta)\mathbf{h}^\Delta + (\partial U/\partial t_0)\mathbf{h}^{n+1}$, the equation (13.12) implies that

$$\frac{-p_\Delta{}^0}{\partial U/\partial q_0{}^\Delta} = \frac{H_0}{\partial U/\partial t_0}, \qquad \Delta = 1, \ldots, n. \tag{13.13}$$

We call (13.13) the system of *transversality conditions*, which is the system of algebraic equations for the generalized momentum $(p_\Delta{}^0)$ at each point $(\dot{q}_0{}^\Delta, t_0)$ in the initial manifold. Assuming that the system (13.13) has a unique solution, we can define a particular generalized momentum $(p_\Delta{}^0)$, and then a particular generalized velocity $(\dot{q}_0{}^\Delta)$ via the Legendre transformation at each point of the initial manifold.

In general, when the initial manifold defined by the condition (13.10), i.e., $U = 0$, is not a Monge strip manifold, the direction of the transversal extremal is not tangent to the initial manifold. Then we can define a function

$$U(q^\Delta, t) = G(q_0{}^\Delta, t_0, q^\Delta, t), \tag{13.14}$$

where the geodetic distance $G(q_0{}^\Delta, t_0, q^\Delta, t)$ is taken along the unique transversal extremal from $(q_0{}^\Delta, t_0)$ in the initial manifold to the point (q^Δ, t) in $\tilde{\mathscr{M}}$. We claim that this function U, like the function V defined earlier by (13.9), is a solution of the Hamilton–Jacobi equation. In fact, we can regard V as a limiting case of U, when the initial manifold shrinks to a single initial point $(q_0{}^\Delta, t_0)$.

The proof that U is a solution of the Hamilton–Jacobi equation follows from (13.3) as in the preceding case. We consider the function U on an arbitrary curve $(q^\Delta(\tau), t(\tau))$. Let $(q_0{}^\Delta(\tau), t_0(\tau))$ be the corresponding curve in the initial manifold such that for each fixed τ the extremal curve joining $(q_0{}^\Delta(\tau), t_0(\tau))$ to $(q^\Delta(\tau), t(\tau))$ is transversal relative to the initial manifold. Then by the definition (13.14) we have

$$U\big(q^\Delta(\tau), t(\tau)\big) = G\big(q_0{}^\Delta(\tau), t_0(\tau), q^\Delta(\tau), t(\tau)\big). \tag{13.15}$$

Now using the general result (13.3), we can calculate the derivative of U with respect to τ by

$$\begin{aligned}\frac{dU}{d\tau} &= \left(p_\Delta \frac{dq^\Delta}{d\tau} - H\frac{dt}{d\tau}\right)\bigg|_{t(\tau)} - \left(p_\Delta \frac{dq^\Delta}{d\tau} - H\frac{dt}{d\tau}\right)\bigg|_{t_0(\tau)} \\ &= \left(p_\Delta \frac{dq^\Delta}{d\tau} - H\frac{dt}{d\tau}\right)\bigg|_{t(\tau)},\end{aligned} \tag{13.16}$$

where the third term vanishes since for a transversal extremal the condition (13.12) holds for any curve $(q_0{}^\Delta(\tau), t_0(\tau))$ in the initial manifold. From (13.16) we can read off as before the special case

$$\frac{\partial U}{\partial q^\Delta} = p_\Delta, \qquad \frac{\partial U}{\partial t} = -H, \tag{13.17}$$

by choosing the curve $(q^\Delta(\tau), t(\tau))$ to be the coordinate curves of (q^Δ, t). Thus U satisfies the Hamilton–Jacobi equation, viz.,

$$\frac{\partial U}{\partial t} + H(q^\Delta, p_\Delta, t) = \frac{\partial U}{\partial t} + H\left(q^\Delta, \frac{\partial U}{\partial q^\Delta}, t\right) = 0. \tag{13.18}$$

Note. The notation $U(q^\Delta, t)$ defined by (13.14) is consistent with the notation $U(q^\Delta, t)$ in the condition (13.10) for the initial manifold, since $U(q^\Delta, t)$ in fact vanishes on the initial manifold. Further, when the initial manifold is nondegenerate, the solution $U(q^\Delta, t)$, unlike the solution $V(q^\Delta, t)$ defined earlier by (13.9) relative to a single initial point $(q_0{}^\Delta, t_0)$, generally has no singularity; i.e., $U(q^\Delta, t)$ is a regular solution of the Hamilton–Jacobi equation.

The converse of the preceding result, i.e., every regular solution $U(q^\Delta, t)$ of the Hamilton–Jacobi equation can be regarded as the geodetic distance along the transversal extremal relative to some initial manifold $U(q^\Delta, t) = \text{const}$, is also valid. We can prove this converse result in the following way: Relative to any given regular solution $U(q^\Delta, t)$ of the Hamilton–Jacobi equation, we define first the generalized momentum p_Δ by

$$p_\Delta = p_\Delta(q^\Gamma, t) = \frac{\partial U}{\partial q^\Delta}, \qquad \Delta = 1, \ldots, n. \tag{13.19}$$

Then from (13.18)

$$\frac{\partial U}{\partial t} = -H(q^\Delta, p_\Delta, t). \tag{13.20}$$

Now we consider the following system of ordinary differential equations:

$$\frac{dq^\Delta}{dt} = \left.\frac{\partial H}{\partial p_\Delta}\right|_{(q^\Gamma, p_\Gamma(q^\Delta, t), t)}, \qquad \Delta = 1, \ldots, n. \tag{13.21}$$

Integrating this system and using the initial manifold $U(q^\Delta, t) = \text{const}$ as the initial condition, we get an n-parameter family of solution curves $(q^\Delta(t), t)$, one for each point in the initial manifold. Substituting these solutions $q^\Delta = q^\Delta(t)$ into (13.19), we obtain the corresponding curves $(q^\Delta(t), p_\Delta(t), t)$ in the phase space $\widetilde{\mathscr{C}^*(\mathscr{M})}$. We claim that these curves are trajectories of the Hamiltonian system (9.10), so that the solution curves $(q^\Delta(t), t)$ are extremal curves of the integral (7.10).

Since the curves $(q^\Delta(t), t)$ are obtained from the solutions of the system (13.21), it suffices to show that

$$\frac{dp_\Delta}{dt} = -\left.\frac{\partial H}{\partial q^\Delta}\right|_{(q^\Gamma(t), p_\Gamma(t), t)}, \tag{13.22}$$

where $p_\Delta(t) = p_\Delta(q^\Gamma(t), t)$. To prove that $p_\Delta(t)$ satisfy the system (13.22), we differentiate the definition (13.19) with respect to t, obtaining

$$\frac{dp_\Delta}{dt} = \frac{\partial^2 U}{\partial q^\Delta\, \partial t} + \frac{\partial^2 U}{\partial q^\Delta\, \partial q^\Gamma}\frac{dq^\Gamma}{dt}, \qquad \Delta = 1, \ldots, n. \tag{13.23}$$

Now since U is a regular solution of the Hamilton–Jacobi equation, we get from differentiating (13.18) the condition

$$0 = \frac{\partial^2 U}{\partial q^\Delta\, \partial t} + \frac{\partial H}{\partial q^\Delta} + \frac{\partial H}{\partial p_\Gamma}\frac{\partial^2 U}{\partial q^\Gamma\, \partial q^\Delta}. \tag{13.24}$$

Combining (13.23) and (13.24) and using (13.21), we obtain (13.22). Thus the n-parameter family of curves $(q^\Delta(t), t)$ defined previously are all extremal curves.

Next we show that the curves $(q^\Delta(t), t)$ are, in fact, transversal extremals relative to the initial manifold; i.e., they satisfy the transversality conditions (13.13) at each point $(q_0{}^\Delta, t_0)$ in the initial manifold. This fact is more or less obvious, since from (13.19) and (13.18)

$$\frac{-p_\Delta}{\partial U/\partial q^\Delta} = \frac{H}{\partial U/\partial t} = -1 \tag{13.25}$$

on all surfaces of the form

$$U(q^\Delta, t) = \text{const.} \tag{13.26}$$

Thus the curves $(q^\Delta(t), t)$ are transversal relative to all these manifolds.

Now along any transversal extremal $(q^\Delta(t), t)$ we have

$$\frac{dU(q^\Delta(t), t)}{dt} = \frac{\partial U}{\partial q^\Delta}\frac{dq^\Delta}{dt} + \frac{\partial U}{\partial t} = p_\Delta \frac{dq^\Delta}{dt} - H = L(q^\Delta, \dot{q}^\Delta, t). \tag{13.27}$$

Consequently, the value of U is simply the geodetic distance measured from the initial manifold, and the proof of the converse result is complete.

The assertion that the curves $(q^\Delta(t), t)$ are transversal relative to all surfaces of the form (13.26) gives rise to an interesting interpretation of the general regular solution U in terms of the special solution V. From the definition of a transversal extremal, we know that the geodetic distance from a fixed point $(q_0{}^\Delta, t_0)$ on $U(q^\Delta, t) = c_0$ to an arbitrary point (q^Δ, t) on $U(q^\Delta, t) = c$ takes on an extremal value $c - c_0$, and this extremal value is achieved along the transversal extremal curve $(q^\Delta(t), t)$ passing through the point $(q_0{}^\Delta, t_0)$. As a result, we can visualize the manifold $U(q^\Delta, t) = c$ as the envelope of the n-parameter family of manifolds

$$G(q_0{}^\Delta, t_0, q^\Delta, t) = c - c_0 \tag{13.28}$$

for all $(q_0{}^\Delta, t_0)$ in the n-dimensional initial manifold $U(q^\Delta, t) = c_0$.

This interpretation corresponds to *Huygens' principle* in optics, which asserts that the wave front $U(q^\Delta, t) = c + \Delta c$ can be visualized as the envelope of the wave fronts $G(q_c{}^\Delta, t_c, q^\Delta, t) = c$ emitting from the point sources at $(q_c{}^\Delta, t_c)$ in the preceding wave front $U(q^\Delta, t) = c$ with an increment of geodetic distance Δc. This general property of the solution of the Hamilton–Jacobi equation is important in the transition from classical mechanics to quantum mechanics.

Appendix. Characteristics of a First-Order Partial Differential Equation

We consider first the partial differential equation

$$\Phi\left(x, y, u, \frac{\partial u}{\partial x}, \frac{\partial u}{\partial y}\right) = 0, \tag{A.1}$$

where $u = u(x, y)$ is the unknown function. The method of characteristics for the equation (A.1) may be summarized as follows:

Suppose that $u = u(x, y)$ is a particular solution surface of (A.1) in the (x, y, u) space. For brevity we denote the partial derivatives $\partial u/\partial x$ and $\partial u/\partial y$ by u_x and u_y, respectively. Then the tangent plane of the surface is perpendicular to the vector $(u_x, u_y, -1)$ in $\mathscr{R}^3$. From (A.1) we see that at each point (x_0, y_0, u_0) the possible values of $u_x(x_0, y_0)$ and $u_y(x_0, y_0)$ must satisfy the algebraic equation

$$\Phi(x_0, y_0, u_0, u_x(x_0, y_0), u_y(x_0, y_0)) = 0, \tag{A.2}$$

which means that there is a functional relation between the x component $u_x(x_0, y_0)$ and the y component $u_y(x_0, y_0)$ of the normal vector $\big(u_x(x_0, y_0), u_y(x_0, y_0), -1\big)$ at the point (x_0, y_0, u_0). Consequently, the possible normals of the solution surface form a cone with vertex at (x_0, y_0, u_0), such that the algebraic equation (A.2) defines the cross section of this cone on the plane $u - u_0 = -1$. Since each possible normal is perpendicular to a possible tangent plane, the family of all possible tangent planes forms the envelope of another cone with vertex at (x_0, y_0, u_0), called the *Monge cone.* We can derive the governing equation for the Monge cone in the following way:

Let $\xi = u_x(x_0, y_0)$ and $\eta = u_y(x_0, y_0)$ be the x component and the y component of a typical normal vector $(\xi, \eta, -1)$ at (x_0, y_0, u_0). Then the equation for the corresponding tangent plane is

$$\xi(x - x_0) + \eta(y - y_0) - (u - u_0) = 0. \tag{A.3}$$

The condition (A.2) shows that ξ and η are functionally related; we can express this relation implicitly by a curve

$$\xi = \xi(\tau), \qquad \eta = \eta(\tau). \tag{A.4}$$

Then the Monge cone is generated by the 1-parameter family of tangent planes with ξ and η given by (A.4). We can determine a typical generator of the Monge cone by the limit of the intersecting lines of adjacent tangent planes. That is, we consider the plane at τ given by

$$\xi(\tau)(x - x_0) + \eta(\tau)(y - y_0) - (u - u_0) = 0 \tag{A.5}$$

and the plane at $\tau + \Delta\tau$ given by

$$\xi(\tau + \Delta\tau)(x - x_0) + \eta(\tau + \Delta\tau)(y - y_0) - (u - u_0) = 0. \tag{A.6}$$

The limit of the intersection, then, satisfies (A.5) and

$$\xi'(\tau)(x - x_0) + \eta'(\tau)(y - y_0) = 0. \tag{A.7}$$

Now from (A.2)

$$\Phi\big(x_0, y_0, u_0, \xi(\tau), \eta(\tau)\big) = 0. \tag{A.8}$$

Differentiating this identity with respect to τ, we obtain

$$\Phi_\xi\big(x_0, y_0, u_0, \xi(\tau), \eta(\tau)\big)\xi'(\tau) + \Phi_\eta\big(x_0, y_0, u_0, \xi(\tau), \eta(\tau)\big)\eta'(\tau) = 0. \tag{A.9}$$

Thus we can replace (A.7) by

$$\frac{x - x_0}{\Phi_\xi} = \frac{y - y_0}{\Phi_\eta}, \tag{A.10}$$

where the arguments of Φ_ξ and Φ_η are shown in (A.9). Combining (A.10) with (A.5), we see that the Monge cone is generated by

$$\frac{x - x_0}{\Phi_\xi} = \frac{y - y_0}{\Phi_\eta} = \frac{u - u_0}{\xi\Phi_\xi + \eta\Phi_\eta}, \tag{A.11}$$

where ξ and η are related by (A.8), which is also given implicitly by (A.4).

Having defined the concept of a Monge cone at each point in the (x, y, u) space, we now consider a curve $(x(\tau), y(\tau), u(\tau))$ whose tangent vector $(x'(\tau), y'(\tau), u'(\tau))$ is a generator of the Monge cone at $(x(\tau), y(\tau), u(\tau))$ for each τ. Such a curve may be assigned a tangent plane of the Monge cone. Let this tangent plane have the normal vector $(\xi(\tau), \eta(\tau), -1)$. Then the five functions $(x(\tau), y(\tau), u(\tau), \xi(\tau), \eta(\tau))$ defined in this way characterize a curve and a tangent plane at each point of the curve. We call such a geometric object a *Monge strip*. From (A.11) and (A.8) the governing equations for a Monge strip are

$$\frac{x'}{\Phi_\xi} = \frac{y'}{\Phi_\eta} = \frac{u'}{\xi\Phi_\xi + \eta\Phi_\eta}, \qquad \Phi(x, y, u, \xi, \eta) = 0. \tag{A.12}$$

In general a Monge strip may or may not be contained in a solution surface $u = u(x, y)$ of the equation (A.1), however. A criterion for a Monge strip to be contained in a solution surface may be derived in the following way: In order that there be a solution $u = u(x, y)$ such that

$$u(\tau) = u(x(\tau), y(\tau)), \tag{A.13a}$$

$$\xi(\tau) = u_x(x(\tau), y(\tau)), \tag{A.13b}$$

$$\eta(\tau) = u_y(x(\tau), y(\tau)) \tag{A.13c}$$

for all τ, it is necessary and sufficient that in addition to (A.12), $\xi(\tau)$ and $\eta(\tau)$ satisfy the *characteristic conditions*:

$$\frac{x'}{\Phi_\xi} = \cdots = \frac{\xi'}{-(\xi\Phi_u + \Phi_x)} = \frac{\eta'}{-(\eta\Phi_u + \Phi_y)}. \tag{A.14}$$

We call a Monge strip satisfying (A.14) a *characteristic strip*.

To prove necessity of (A.14), we obtain first from (A.13b)

$$\xi' = u_{xx}x' + u_{xy}y'. \tag{A.15}$$

Next, differentiating (A.12c) with respect to x, we get

$$\Phi_x + \Phi_u u_x + \Phi_\xi u_{xx} + \Phi_\eta u_{xy} = 0. \tag{A.16}$$

Evaluating this relation on the strip and using (A.15) and (A.12a), we obtain

$$\Phi_x + \Phi_u \xi + \frac{\Phi_\xi}{x'}\,\xi' = 0, \tag{A.17}$$

which can be rewritten as

$$\frac{x'}{\Phi_\xi} = \frac{\xi'}{-(\xi\Phi_u + \Phi_x)}. \tag{A.18}$$

The same argument can be applied to y and η. Thus (A.14) is necessary for a Monge strip to be a strip of a solution surface.

Notice that the common ratio in (A.14) is a function of the point of the strip, i.e., a function of τ. Hence after an appropriate reparametrization (A.14) may be reduced to the system of *characteristic equations*:

$$\begin{aligned} x' = \Phi_\xi, \quad y' = \Phi_\eta, \quad u' = \xi\Phi_\xi + \eta\Phi_\eta, \quad \xi' = -\xi\Phi_u - \Phi_x, \\ \eta' = -\eta\Phi_u - \Phi_y, \end{aligned} \tag{A.19}$$

for the five coordinates (x, y, u, ξ, η) of the characteristic strip. We can prove that the function Φ is an integral of the system (A.19), viz.,

$$\begin{aligned} \frac{d\Phi}{d\tau} &= \Phi_x x' + \Phi_y y' + \Phi_u u' + \Phi_\xi \xi' + \Phi_\eta \eta' \\ &= \Phi_x\Phi_\xi + \Phi_y\Phi_\eta + \Phi_u(\xi\Phi_\xi + \eta\Phi_\eta) - \Phi_\xi(\xi\Phi_u + \Phi_x) - \Phi_\eta(\eta\Phi_u + \Phi_y) \\ &= 0. \end{aligned} \tag{A.20}$$

Consequently, if $\Phi = 0$ at an initial point of the strip, then the solution of the system (A.19) automatically satisfies the condition (A.12c).

Now we show that, conversely, every strip satisfying (A.19) and the condition $\Phi = 0$ is contained in a solution surface of the partial differential equation (A.1). This result is known as the *method of characteristics* for the construction of a regular solution of (A.1).

In general any strip $(x(t), y(t), u(t), \xi(t), \eta(t))$ which belongs to a solution surface $u = u(x, y)$ of (A.1) clearly must satisfy the conditions

$$\dot{u} = \xi\dot{x} + \eta\dot{y}, \qquad \Phi(x, y, u, \xi, \eta) = 0 \tag{A.21}$$

for all t, where the superposed dot denotes the derivative with respect to the parameter t. We call any strip satisfying (A.21) an *initial strip*. This term is suggested by the fact that each point $(x(t), y(t), u(t), \xi(t), \eta(t))$ in an initial strip may be regarded as an initial point for the system (A.19). Using the initial strip as a strip of initial points, we can solve the autonomous system (A.19) and obtain a 1-parameter family of characteristic strips of the form

$$x = x(t, \tau), \quad y = y(t, \tau), \quad u = u(t, \tau), \quad \xi = \xi(t, \tau), \quad \eta = \eta(t, \tau). \tag{A.22}$$

Now there are two possibilities for this 2-parameter family of points:

(i) The transformation $x = x(t, \tau)$, $y = y(t, \tau)$ is invertible at the initial strip $\tau = 0$. In this case the initial strip is contained in a unique solution surface $u = u(x, y)$ which may be obtained by inverting the transformation and substituting the result into the function $u = u(t, \tau)$.

To prove this assertion we observe first that since (A.22) is obtained from (A.19), by the initial condition (A.21) and the integral condition (A.20)

$$\Phi(x(t, \tau), y(t, \tau), u(t, \tau), \xi(t, \tau), \eta(t, \tau)) = 0 \tag{A.23}$$

for all t and τ. Now by the chain rule the surface $u = u(x, y)$ obtained from the transformation satisfies the relations

$$u_t = u_x x_t + u_y y_t, \qquad u_\tau = u_x x_\tau + u_y y_\tau. \tag{A.24}$$

We claim that the same relations hold when u_x and u_y are replaced by ξ and η, viz.,

$$u_t = \xi x_t + \eta y_t, \tag{A.25a}$$

$$u_\tau = \xi x_\tau + \eta y_\tau. \tag{A.25b}$$

Notice that the relation (A.25a) holds initially at $\tau = 0$; cf. (A.19c).

To prove that (A.25a) hold for all τ, we define

$$\zeta = \zeta(t, \tau) = (u_t - \xi x_t - \eta y_t)(t, \tau). \tag{A.26}$$

Then $\zeta(t, 0) = 0$ initially. We claim that ζ satisfies the linear ordinary differential equation (τ is the independent variable and t is the parameter)

$$\zeta_\tau = -\Phi_u \zeta \tag{A.27}$$

for each t. We can prove this result by direct differentiation of (A.26), obtaining first

$$\zeta_\tau = u_{t\tau} - \xi_\tau x_t - \xi x_{t\tau} - \eta_\tau y_t - \eta y_{t\tau}. \tag{A.28}$$

Now differentiating (A.25b) with respect to t, we get

$$0 = u_{t\tau} - \xi_t x_\tau - \xi x_{t\tau} - \eta_t y_\tau - \eta y_{t\tau}. \tag{A.29}$$

Subtracting (A.29) from (A.28) and then using (A.19), we get

$$\begin{aligned}\zeta_\tau &= -\xi_\tau x_t + \xi_t x_\tau - \eta_\tau y_t + \eta_t y_\tau \\ &= (\xi\Phi_u + \Phi_x)x_t + \Phi_\xi \xi_t + (\eta\Phi_u + \Phi_y)y_t + \Phi_\eta \eta_t \\ &= \Phi_x x_t + \Phi_y y_t + \Phi_\xi \xi_t + \Phi_\eta \eta_t + (\xi x_t + \eta y_t)\Phi_u.\end{aligned} \tag{A.30}$$

From (A.23) we have also

$$0 = \Phi_x x_t + \Phi_y y_t + \Phi_\xi \xi_t + \Phi_\eta \eta_t + \Phi_u u_t. \tag{A.31}$$

Combining (A.30) and (A.31) and using the definition (A.26), we obtain (A.27). As a result, ζ must vanish for all τ, since it vanishes initially. Thus (A.25) is proved.

Now under the assumption of invertibility the Jacobian matrix $[\partial(x, y)/\partial(t, \tau)]$ is nonsingular. Hence (A.24) and (A.25) imply that $\xi = u_x$ and $\eta = u_y$. Thus from (A.23) we see that $u = u(x, y)$ is a solution of the partial differential equation (A.1).

The assertion which we have just established is the central idea of the method of characteristics; it shows that a regular solution of the partial differential equation (A.1) may be obtained by a 1-parameter family of characteristic strips over a (noncharacteristic) initial strip. If the initial strip is itself a characteristic strip, then the case belongs to the next possibility:

(ii) The transformation $x = x(t, \tau)$, $y = y(t, \tau)$ is not invertible at the initial strip $\tau = 0$. In this case the initial strip must be a Monge strip.

To prove this assertion notice that the condition $\det[\partial(x, y)/\partial(t, \tau)] = 0$ at $\tau = 0$ implies that the initial strip satisfies

$$0 = (x_t y_\tau - x_\tau y_t)\big|_{\tau=0} = (\dot{x}\Phi_\eta - \dot{y}\Phi_\xi)\big|_{\tau=0} \tag{A.32}$$

or, equivalently, on $(x(t), y(t), u(t), \xi(t), \eta(t))$

$$\frac{\dot{x}}{\Phi_\xi} = \frac{\dot{y}}{\Phi_\eta}. \tag{A.33}$$

Combining this with the condition (A.21), we see that the initial strip is a Monge strip with parameter t. Hence there are two possibilities:

(iia) The initial strip is a characteristic strip. In this case there are infinitely many solution surfaces containing the initial strip.

For comparison with case (i) we change first the parameter of the initial (characteristic) strip to τ. Now at $\tau = 0$ we attach a matching initial strip which is not a Monge strip. Aside from a single point at $\tau = 0$, the initial strip is entirely arbitrary. For each such initial strip we can generate a solution surface as in case (i) by a 1-parameter family of characteristic strips which includes the given characteristic strip. This result now shows that (A.14) is sufficient for a Monge strip to be a strip of a solution surface.

(iib) The initial strip is a noncharacteristic Monge strip. In this case in proving the necessity of (A.14) we have shown that there is no solution surface containing the given initial strip.

The method of characteristics summarized above for the simple case of first-order partial differential equation (A.1) of two independent variables (x, y) may be generalized to first-order partial differential equations in more than two independent variables, say an equation of the form

$$\Phi(x^i, u, u_{x^i}) = \Phi\left(x^1, \ldots, x^m, u, \frac{\partial u}{\partial x^1}, \ldots, \frac{\partial u}{\partial x^m}\right) = 0 \tag{A.34}$$

in m independent variables. As before we put $u_{x^i} = \xi_i$, $i = 1, \ldots, m$. Then at any point $(x_0{}^i, u_0) \in \mathscr{R}^{m+1}$ we define the *Monge cone* by

$$\frac{x^1 - x_0{}^1}{\Phi_{\xi_1}} = \cdots = \frac{x^m - x_0{}^m}{\Phi_{\xi_m}} = \frac{u - u_0}{\xi_i \Phi_{\xi_i}}, \tag{A.35}$$

where the repeated index i is summed from 1 to m. Similarly, we define a Monge strip $(x^i(\tau), u(\tau), \xi_i(\tau))$ by the conditions

$$\frac{(x^1)'}{\Phi_{\xi_1}} = \cdots = \frac{(x^m)'}{\Phi_{\xi_m}} = \frac{u'}{\xi_i \Phi_{\xi_i}}, \qquad \Phi(x^i, u, \xi_i) = 0. \tag{A.36}$$

A Monge strip is contained in a solution surface $u = u(x^i)$ if and only if it is a characteristic strip which is defined by the *characteristic conditions*

$$\frac{(x^i)'}{\Phi_{\xi_i}} = \frac{u'}{\xi_j \Phi_{\xi_j}} = \frac{(\xi_i)'}{-(\xi_i \Phi_u + \Phi_{x^i})}, \qquad \Phi(x^k, u, \xi_k) = 0, \qquad \text{(A.37)}$$

where $i = 1, \ldots, m$, and where the repeated index j is summed from 1 to m.

As before we can rewrite (A.37) as the system of *characteristic equations*

$$(x^i)' = \Phi_{\xi_i}, \qquad u' = \xi_j \Phi_{\xi_j}, \qquad (\xi_i)' = -(\xi_i \Phi_u + \Phi_{x^i}), \qquad \text{(A.38)}$$

and the function $\Phi(x^k, u, \xi_k)$ is an integral of this system. Consequently, we can determine a characteristic strip by solving the system (A.38) with an initial point $\left(x^i(0), u(0), \xi_i(0)\right)$ satisfying the algebraic condition $\Phi = 0$.

The concept of an initial strip for an equation of two independent variables can be generalized to an *initial strip manifold* of dimension $m - 1$ given by

$$x^i = x^i(t^1, \ldots, t^{m-1}), \qquad u = u(t^\alpha), \qquad \xi_i = \xi_i(t^\alpha), \qquad \text{(A.39)}$$

where α ranges from 1 to $m - 1$. The coordinate functions of an initial strip manifold are required to satisfy the following *initial conditions*:

$$u_{t^\alpha} = \xi_i x^i_{t^\alpha}, \qquad \alpha = 1, \ldots, m - 1, \qquad \Phi(x^k, u, \xi_k) = 0, \qquad \text{(A.40)}$$

at all points of the manifold. These conditions are necessary in order that the initial strip manifold be contained in a solution surface.

We call an initial strip manifold a *Monge strip manifold* if it is formed by Monge strips. That is, we can find a new coordinate system $(\bar{t}^1, \ldots, \bar{t}^{m-2}, \tau)$, which is related to the coordinate system $(t^1, \ldots, t^{m-1})$ by a coordinate transformation

$$\bar{t}^1 = \bar{t}^1(t^\alpha), \ldots, \bar{t}^{m-2} = \bar{t}^{m-2}(t^\alpha), \qquad \tau = \tau(t^\alpha), \qquad \text{(A.41)}$$

such that the coordinate strips of τ are all Monge strips, viz., they satisfy (A.36), or, equivalently,

$$x_\tau^i = \Phi_{\xi_i}, \qquad u_\tau = \xi_j \Phi_{\xi_j}, \qquad \Phi(x^i, u, \xi_i) = 0. \qquad \text{(A.42)}$$

A Monge strip manifold can be visualized simply as an $(m - 2)$-parameter family of Monge strips generated by an initial strip manifold of dimension $(m - 2)$. This lower-dimensional initial strip manifold must not be a Monge strip manifold itself, of course; otherwise, the Monge strips generated from it will stay in the initial strip manifold.

We call an initial strip manifold a *characteristic strip manifold* if it is formed by characteristic strips. In this case there is a new coordinate system $(\bar{t}^1, \ldots, \bar{t}^{m-2}, \tau)$ such that the τ coordinate strips are all characteristic strips, i.e., they satisfy (A.38) with respect to τ for each $(\bar{t}^1, \ldots, \bar{t}^{m-2})$. We can visualize a characteristic strip manifold simply as an $(m-2)$-parameter family of characteristic strips generated by a noncharacteristic initial strip manifold of dimension $(m-2)$.

Note. We can define the concepts of initial strip manifold, Monge strip manifold, and characteristic strip manifold of dimension lower than $(m-1)$ in a similar way. However, these lower-dimensional manifolds are not important in the description of a solution surface, which is just a characteristic strip manifold of dimension m, as we shall now see.

We can visualize the solution surface of (A.34) for a given initial strip manifold in the following way: Starting from the initial points (A.39), we solve the problem (A.38) and obtain the solution

$$x^i = x^i(t^\alpha, \tau), \qquad u = u(t^\alpha, \tau), \qquad \xi_i = \xi_i(t^\alpha, \tau). \tag{A.43}$$

As before there are two possibilities:

(i) The transformation $x^i = x^i(t^\alpha, \tau)$ is invertible at the initial strip manifold. In this case the initial strip manifold is contained in a unique solution surface $u = u(x^i)$, which may be obtained by inverting the transformation and substituting the result into the function $u = u(t^\alpha, \tau)$.

(ii) The transformation $x^i = x^i(t^\alpha, \tau)$ is not invertible at the initial strip manifold. In this case the initial strip manifold must be a Monge strip manifold. Again there are two possibilities:

(iia) The initial strip manifold is a characteristic strip manifold. In this case there are infinitely many solution surfaces containing the initial strip manifold.

(iib) The initial strip manifold is a noncharacteristic Monge strip manifold. In this case there is no solution surface containing the initial strip manifold.

The results summarized above can be proven in the same way as their counterparts in the case of two independent variables.

3

Basic Principles of Continuum Mechanics

Continuum mechanics is the branch of classical mechanics concerned with motions of deformable material bodies. The mathematical model for such a body is called a *body manifold* which is an oriented 3-dimensional differentiable manifold endowed with global coordinate systems. In this chapter we develop the basic principles governing the motions of body manifolds.

14. Deformations and Motions

The concepts of Newtonian space–time and a frame of reference introduced in Section 1 are basic notions in classical mechanics. Since continuum mechanics is one of its subjects, these notions remain applicable in this chapter. We choose a particular frame of reference, and we denote the corresponding physical space by $\mathscr{S}$. To specify a position $\mathbf{x}$ in $\mathscr{S}$, we use a Cartesian coordinate system (x^i); i.e., we locate $\mathbf{x}$ by

$$\mathbf{x} = \mathbf{o} + x^i \mathbf{e}_i, \tag{14.1}$$

where $\mathbf{o}$ and $\{\mathbf{e}_i\}$ denote the origin and the coordinate basis of the Cartesian system. Relative to that coordinate system $\mathscr{S}$ is represented isometrically by the oriented inner product space $\mathscr{R}^3$.

The physical objects considered by continuum mechanics are deformable material bodies which are represented mathematically by various body manifolds. We define a *body manifold* $\mathscr{B}$ as follows: an oriented 3-dimensional differentiable manifold which is endowed with global coordinate systems. Using a Cartesian system (x^i) on $\mathscr{S}$, we may regard a global coordinate system on $\mathscr{B}$, say,

$$\boldsymbol{\chi}: \mathscr{B} \to \mathscr{R}^3 \xleftrightarrow{(x^i)} \mathscr{S}, \tag{14.2}$$

as an (orientation-preserving) diffeomorphism of $\mathscr{B}$ into $\mathscr{S}$. Such a diffeomorphism is called a *configuration* of $\mathscr{B}$.

Note. A configuration of a body manifold is a concept similar to a configuration of a rigid body, except that, here, $\boldsymbol{\chi}$ is only required to be a diffeomorphism, not necessarily an isometry. In other words, a body manifold may *deform* from one configuration to another one. This distinction is a major generalization from the theory of rigid bodies, since unlike an isometry, a diffeomorphism in general cannot be characterized by a finite set of parameters.

To specify a configuration $\boldsymbol{\chi}$ of $\mathscr{B}$, we assign the coordinates (x^i) of the position $\mathbf{x} = \boldsymbol{\chi}(X) \in \mathscr{S}$ occupied by each body point $X \in \mathscr{B}$. Since one global coordinate system of $\mathscr{B}$ identifies uniquely the entirety of all coordinate systems of $\mathscr{B}$ via the family of coordinate transformations, we may use a particular reference configuration $\boldsymbol{\varkappa}$ to characterize $\mathscr{B}$. It is customary to denote the Cartesian coordinates of $\boldsymbol{\varkappa}(X)$ by (X^A); i.e.,

$$\boldsymbol{\varkappa}: \mathscr{B} \to \mathscr{R}^3, \qquad \boldsymbol{\varkappa}(X) = (X^A) = \big(X^A(X)\big). \tag{14.3}$$

We call $\big(X^A(X)\big)$ the *referential coordinates* of X.

A configuration $\boldsymbol{\chi}$ in general may be characterized by a coordinate transformation from $\boldsymbol{\varkappa}$ to $\boldsymbol{\chi}$, viz.,

$$\boldsymbol{\varphi} \equiv \boldsymbol{\chi} \circ \boldsymbol{\varkappa}^{-1}: \boldsymbol{\varkappa}(\mathscr{B}) \to \boldsymbol{\chi}(\mathscr{B}), \qquad x^i = x^i(X^A), \qquad i = 1, 2, 3. \tag{14.4}$$

The mapping $\boldsymbol{\varphi}$ is the *deformation* from the configuration $\boldsymbol{\varkappa}$ to the configuration $\boldsymbol{\chi}$, so we call $x^i(X^A)$, $i = 1, 2, 3$, the *deformation functions.*

If we identify the body manifold $\mathscr{B}$ with the domain $\boldsymbol{\varkappa}(\mathscr{B})$ in $\mathscr{S}$, then the tangent space $\mathscr{B}_X$ of $\mathscr{B}$ at any body point $X \in \mathscr{B}$ is just the physical translation space $\mathscr{V}$. Indeed, the tangent vector $\mathbf{h}_A$ of any coordinate line of X^A coincides with the basis vector $\mathbf{e}_A$. Hence the natural basis $\{\mathbf{h}_A\}$ of (X^A) at any body point $X \in \mathscr{B}$ is the same as the coordinate basis $\{\mathbf{e}_A\}$ of the Cartesian system on $\mathscr{S}$.

Now consider a deformation $\boldsymbol{\varphi}$ from $\boldsymbol{\varkappa}$ to $\boldsymbol{\chi}$. From (14.4) a coordinate line of X^A in $\boldsymbol{\varkappa}(\mathscr{B})$ is mapped into a curve with coordinates $x^i = x^i(X^A)$

in $\boldsymbol{\chi}(\mathscr{B})$, and the tangent vector of this curve is

$$\frac{\partial x^i}{\partial X^A}\,\mathbf{e}_i = [\text{grad}\,\boldsymbol{\varphi}](\mathbf{e}_A) \equiv \mathbf{F}\mathbf{e}_A. \tag{14.5}$$

We call the field of linear maps

$$\mathbf{F}(\mathbf{X})\colon \mathscr{V} \to \mathscr{V}, \qquad \mathbf{X} \in \boldsymbol{\varkappa}(\mathscr{B}) \tag{14.6}$$

defined by (14.5) the field of *deformation gradients*. From (14.5) $\mathbf{F}$ has the component form

$$\mathbf{F} = \frac{\partial x^i}{\partial X^A}\,\mathbf{e}_i \otimes \mathbf{e}^A, \tag{14.7}$$

where $\{\mathbf{e}^A\}$ denotes the dual basis of $\{\mathbf{e}_A\}$. (In fact we can identify $\{\mathbf{e}^A\}$ with $\{\mathbf{e}_A\}$, since $\{\mathbf{e}_A\}$ is an orthonormal basis; cf., the explanation given in Section 31, IVT-1.) Since the deformation $\boldsymbol{\varphi}$ is an orientation-preserving diffeomorphism, the determinant of $\mathbf{F}$ is necessarily positive,

$$\det \mathbf{F} = \det\left[\frac{\partial x^i}{\partial X^A}\right] > 0. \tag{14.8}$$

Note. The field of deformation gradients, by its very definition, must satisfy certain compatibility conditions. Specifically, the identities

$$\frac{\partial F_A{}^i}{\partial X^B} = \frac{\partial F_B{}^i}{\partial X^A}, \qquad A, B = 1, 2, 3, \tag{14.9}$$

must hold regardless of the deformation $\boldsymbol{\varphi}$. However, these identities do not restrict the value $\mathbf{F}(\mathbf{X}_0)$ at any one point $\mathbf{X}_0 \in \boldsymbol{\varkappa}(\mathscr{B})$. In fact any matrix $[F_A{}^i]$ which satisfies the condition (14.8) may be the component matrix of the deformation gradient of a homogeneous deformation given by

$$x^i = F_A{}^i X^A, \qquad i = 1, 2, 3. \tag{14.10}$$

In this special case the deformation gradient is a constant field, and the compatibility conditions (14.9) are automatically satisfied.

In Section 27, IVT-1, we showed that every orientation-preserving isomorphism of an inner product space may be regarded as the composition of a positive definite symmetric map and a rotation. This result is known as the polar decomposition theorem. Applying this result to the deformation gradients at all points, we get the field equation

$$\mathbf{F} = \mathbf{RU} = \mathbf{VR}, \tag{14.11}$$

where $\mathbf{U}, \mathbf{V}$ are positive definite symmetric, and where $\mathbf{R}$ is a rotation. We called $\mathbf{U}$ (respectively, $\mathbf{V}$) the field of *right stretch tensors* (respectively, *left stretch tensors*) and $\mathbf{R}$ the field of *rotation tensors* of the deformation $\boldsymbol{\varphi}$.

Since the decomposition (14.11) is defined for each point $\mathbf{X} \in \boldsymbol{\varkappa}(\mathscr{B})$, we can illustrate its meaning by using the special case (14.10) of a homogeneous deformation. In this case the fields $\mathbf{U}$, $\mathbf{V}$, and $\mathbf{R}$, like the field $\mathbf{F}$, are all constant. Then (14.11) means that there are three mutually orthogonal vectors, say $\mathbf{f}_i$, $i = 1, 2, 3$, which are proper vectors of $\mathbf{U}$, such that $\mathbf{Ff}_i$, $i = 1, 2, 3$, remain mutually orthogonal after the deformation is applied to them.

For a general inhomogeneous deformation the preceding interpretation is valid at each point $\mathbf{X}_0 \in \boldsymbol{\varkappa}(\mathscr{B})$. Specifically, there are three mutually perpendicular curves, say $\boldsymbol{\lambda}_i(t)$, $i = 1, 2, 3$, passing through $\mathbf{X}_0$, such that the image curves $\boldsymbol{\varphi}(\boldsymbol{\lambda}_i(t))$, $i = 1, 2, 3$, remain mutually perpendicular at $\mathbf{x}_0 = \boldsymbol{\varphi}(\mathbf{X}_0) \in \boldsymbol{\chi}(\mathscr{B})$. In fact the rotation $\mathbf{R}(\mathbf{X}_0)$ is just the linear map which transforms the unit tangent vectors of $\boldsymbol{\lambda}_i(t)$ at $\mathbf{X}_0$ to the corresponding unit tangent vectors of $\boldsymbol{\varphi}(\boldsymbol{\lambda}_i(t))$ at $\mathbf{x}_0$. Also, the tangent vectors $\dot{\boldsymbol{\lambda}}_i\,|_{\mathbf{X}_0}$ are proper vectors of $\mathbf{U}(\mathbf{X}_0)$, and the tangent vectors

$$(\boldsymbol{\varphi} \circ \dot{\boldsymbol{\lambda}}_i)\,|_{\mathbf{x}_0} = \mathbf{F}(\mathbf{X}_0)\dot{\boldsymbol{\lambda}}_i\,|_{\mathbf{X}_0}, \qquad i = 1, 2, 3, \tag{14.12}$$

are proper vectors of $\mathbf{V}(\mathbf{X}_0)$, the proper numbers α_i associated with $\dot{\boldsymbol{\lambda}}_i\,|_{\mathbf{X}_0}$ and $\mathbf{F}(\mathbf{X}_0)\dot{\boldsymbol{\lambda}}_i\,|_{\mathbf{X}_0}$ being given by

$$\alpha_i = \frac{\|\,\mathbf{F}(\mathbf{X}_0)\dot{\boldsymbol{\lambda}}_i\,|_{\mathbf{X}_0}\,\|}{\|\,\dot{\boldsymbol{\lambda}}_i\,|_{\mathbf{X}_0}\,\|}, \qquad i = 1, 2, 3. \tag{14.13}$$

Note. Although there is a principal basis for $\mathbf{U}(\mathbf{X})$ at each point $\mathbf{X} \in \boldsymbol{\varkappa}(\mathscr{B})$, in general a field of such bases need not be a holonomic field. Hence there may or may not be an orthogonal coordinate net in $\boldsymbol{\varkappa}(\mathscr{B})$ which is mapped to an orthogonal coordinate net in $\boldsymbol{\chi}(\mathscr{B})$ by the deformation $\boldsymbol{\varphi}$.

Having considered the concept of a deformation, we now turn our attention to the more general concept of a *motion* of $\mathscr{B}$. As usual we define a motion as follows: a 1-parameter family of configurations, $\boldsymbol{\chi}_t$, $t \in \mathscr{R}$, where t denotes the time. Relative to a fixed reference configuration $\boldsymbol{\varkappa}$, a motion may be represented by a 1-parameter family of deformations,

$$\boldsymbol{\varphi}_t = \boldsymbol{\chi}_t \circ \boldsymbol{\varkappa}^{-1} : \boldsymbol{\varkappa}(\mathscr{B}) \to \boldsymbol{\chi}_t(\mathscr{B}), \qquad x^i = x^i(X^A, t). \tag{14.14}$$

At each time t we define the field of deformation gradients by taking the partial derivatives of $x^i(X^A, t)$ with respect to X^A, and we define the field of velocities $\mathbf{v}$ and the field of accelerations $\mathbf{a}$ by taking the partial derivatives of $x^i(X^A, t)$ with respect to t, viz.,

$$\mathbf{v} = \frac{\partial x^i}{\partial t}\,\mathbf{e}_i, \qquad \mathbf{a} = \frac{\partial^2 x^i}{\partial t^2}\,\mathbf{e}_i. \tag{14.15}$$

These definitions are consistent with similar definitions given before for particles and rigid bodies, since the time derivatives are taken along the path of each body point X, which is characterized by the referential coordinates (X^A).

In continuum mechanics it is often more convenient to regard $\mathbf{v}$ and $\mathbf{a}$ as vector fields defined on the instantaneous configuration $\boldsymbol{\chi}_t(\mathscr{B})$ rather than as fields on the reference configuration $\boldsymbol{\varkappa}(\mathscr{B})$. This transformation is defined by using the inverse $\boldsymbol{\varphi}_t^{-1}$ of the diffeomorphism $\boldsymbol{\varphi}_t$; i.e.,

$$v^i = v^i(x^j, t) = \frac{\partial x^i}{\partial t}\,(X^A(x^j, t), t), \tag{14.16}$$

and similarly,

$$a^i = a^i(x^j, t) = \frac{\partial^2 x^i}{\partial t^2}\,(X^A(x^j, t), t), \tag{14.17}$$

where $X^A = X^A(x^j, t)$ denotes the inverse of $x^i = x^i(X^A, t)$. Using the chain rule, we can calculate the acceleration field $\mathbf{a}(\mathbf{x}, t)$ from the velocity field $\mathbf{v}(\mathbf{x}, t)$ by *Euler's formula*:

$$a^i = \frac{\partial v^i}{\partial t} + \frac{\partial v^i}{\partial x^j}\,v^j, \qquad i = 1, 2, 3, \tag{14.18}$$

where the independent variables are (x^k, t).

Note. In general we can regard any function $\Phi = \Phi(X^A, t)$ on $\boldsymbol{\varkappa}(\mathscr{B})$ as a function $\Phi = \Psi(x^i, t) = \Phi\big(X^A(x^i, t), t\big)$ on $\boldsymbol{\chi}_t(\mathscr{B})$ via the change of coordinates $x^i = x^i(X^A, t)$ due to the motion $\boldsymbol{\varphi}_t$. Then the time derivative of Φ can be calculated from the representation Ψ by *Euler's formula* in general:

$$\frac{\partial \Phi}{\partial t} = \frac{\partial \Psi}{\partial t} + \frac{\partial \Psi}{\partial x^i}\,v^i. \tag{14.19}$$

It is customary to call the time derivative along the path of a body point (i.e., holding X^A fixed) the *material derivative*. This time derivative should not be confused with the partial derivative with respect to t at a fixed position x^i. A standard notation for the material derivative is a superposed dot. Thus we have $\mathbf{v} = \dot{\mathbf{x}}$, $\mathbf{a} = \dot{\mathbf{v}}$, etc.

The partial derivative $\partial v^i/\partial x^j$, which appears in the last term of (14.18), is known as the *velocity gradient* and is often denoted by the direct tensor notation:

$$\operatorname{grad} \mathbf{v} = \frac{\partial v^i}{\partial x^j} \mathbf{e}_i \otimes \mathbf{e}^j. \tag{14.20}$$

This tensor field may be calculated from the deformation gradient $\mathbf{F}$ by the formula

$$\operatorname{grad} \mathbf{v} = \dot{\mathbf{F}}\mathbf{F}^{-1}, \qquad \frac{\partial v^i}{\partial x^j} = \frac{\partial}{\partial t}\left(\frac{\partial x^i}{\partial X^A}\right)\frac{\partial X^A}{\partial x^j}, \tag{14.21}$$

where $\dot{\mathbf{F}}$ denotes the material derivative of $\mathbf{F}$. In particular, when the configuration $\boldsymbol{\chi}_{t_0}$ at a certain time t_0 is taken to be the reference configuration, the formula (14.21a) reduces to

$$\operatorname{grad} \mathbf{v}(\cdot, t_0) = \dot{\mathbf{F}}(\cdot, t_0), \qquad (\boldsymbol{\varkappa} = \boldsymbol{\chi}_{t_0}) \tag{14.22}$$

since in this case we have

$$\mathbf{F}(\cdot, t_0) = \mathbf{I} \qquad (\boldsymbol{\varkappa} = \boldsymbol{\chi}_{t_0}). \tag{14.23}$$

We call the deformation gradient from $\boldsymbol{\chi}_{t_0}$ to $\boldsymbol{\chi}_t$ the *relative deformation gradient* of the motion from t_0 to t. Then (14.22) shows that the velocity gradient at t_0 is just the material derivative of the deformation gradient relative to $\boldsymbol{\chi}_{t_0}$.

Taking the material derivative of the polar decomposition of $\mathbf{F}$, we get

$$\dot{\mathbf{F}} = \dot{\mathbf{R}}\mathbf{U} + \mathbf{R}\dot{\mathbf{U}} = \dot{\mathbf{V}}\mathbf{R} + \mathbf{V}\dot{\mathbf{R}}. \tag{14.24}$$

Substituting the result into (14.21a) , we obtain

$$\operatorname{grad} \mathbf{v} = \dot{\mathbf{R}}\mathbf{R}^T + \mathbf{R}\dot{\mathbf{U}}\mathbf{U}^{-1}\mathbf{R}^T = \dot{\mathbf{V}}\mathbf{V}^{-1} + \mathbf{V}\dot{\mathbf{R}}\mathbf{R}^T\mathbf{V}^{-1}. \tag{14.25}$$

In particular, when the configuration $\boldsymbol{\chi}_{t_0}$ is taken to be the reference configuration as before, (14.25) reduces to

$$\operatorname{grad} \mathbf{v}(\cdot, t_0) = \dot{\mathbf{R}}(\cdot, t_0) + \dot{\mathbf{U}}(\cdot, t_0) = \dot{\mathbf{V}}(\cdot, t_0) + \dot{\mathbf{R}}(\cdot, t_0), \quad (\boldsymbol{\varkappa} = \boldsymbol{\chi}_{t_0}) \tag{14.26}$$

since from (14.23) we have

$$\mathbf{U}(\cdot, t_0) = \mathbf{V}(\cdot, t_0) = \mathbf{R}(\cdot, t_0) = \mathbf{I} \qquad (\boldsymbol{\varkappa} = \boldsymbol{\chi}_{t_0}). \tag{14.27}$$

The formula (14.26) gives an interesting decomposition for the velocity gradient. Indeed, when (14.27) holds, $\dot{\mathbf{R}}(\cdot, t_0)$ is skew symmetric, and $\dot{\mathbf{U}}(\cdot, t_0) = \dot{\mathbf{V}}(\cdot, t_0)$ is symmetric. Hence (14.26) is just the decomposition of grad $\mathbf{v}(\cdot, t_0)$ into its symmetric part and its skew-symmetric part; i.e.,

$$\dot{\mathbf{R}}(\cdot, t_0) = \tfrac{1}{2}[\text{grad}\, \mathbf{v} - (\text{grad}\, \mathbf{v})^T](\cdot, t_0) \equiv \mathbf{W}(\cdot, t_0), \qquad (\boldsymbol{\varkappa} = \boldsymbol{\chi}_{t_0}) \tag{14.28}$$

and

$$\dot{\mathbf{U}}(\cdot, t_0) = \dot{\mathbf{V}}(\cdot, t_0) = \tfrac{1}{2}[\text{grad}\, \mathbf{v} + (\text{grad}\, \mathbf{v})^T](\cdot, t_0) \equiv \mathbf{D}(\cdot, t_0), \qquad (\boldsymbol{\varkappa} = \boldsymbol{\chi}_{t_0}) \tag{14.29}$$

where $\mathbf{W}$ and $\mathbf{D}$ are called the *spin tensor* and the *stretching tensor*, respectively. Substituting (14.28) and (14.29) into (14.26), we get

$$\text{grad}\, \mathbf{v} = \mathbf{D} + \mathbf{W}, \tag{14.30}$$

which is known as the *Cauchy–Stokes decomposition* of the velocity gradient.

The deformation gradient, the velocity, and the acceleration fields are the basic kinematical quantities of the motion of a body manifold. In the following section we shall define some other kinematical quantities, such as momentum and energy, from these basic quantities. Then we consider the basic balance principles governing the motions of a body manifold.

15. Balance Principles

In Section 2 we defined the concept of a *mass measure* on a rigid body; a similar concept can be defined for a deformable body. Let the body manifold $\mathscr{B}$ be represented by a reference configuration $\boldsymbol{\varkappa}$. Then we define a mass measure μ on $\mathscr{B}$ by

$$\mu(\mathscr{P}) = \int_{\mathscr{P}} d\mu = \int_{\boldsymbol{\varkappa}(\mathscr{P})} \varrho_{\boldsymbol{\varkappa}}\, dX, \tag{15.1}$$

where $\mathscr{P}$ is any subbody of $\mathscr{B}$, and where $\varrho_{\boldsymbol{\varkappa}} = \varrho_{\boldsymbol{\varkappa}}(X^A)$ is a positive (mass) density function on $\boldsymbol{\varkappa}(\mathscr{B})$.

Note. The representation of μ by an integral over $\boldsymbol{\varkappa}(\mathscr{B})$ requires an assumption that the mass measure be absolutely continuous with respect to the Euclidean volume measure on $\boldsymbol{\varkappa}(\mathscr{B})$; i.e., when the volume of $\boldsymbol{\varkappa}(\mathscr{P})$ approaches zero, the mass $\mu(\mathscr{P})$ must also. In particular, this assumption rules out the existence of any concentrated mass particle in a body manifold. The reason for accepting this assumption is because continuum mechanics generally deals with distributed force fields on configurations. Any concentrated mass point would require a concentrated force in its motion and thus cause a singularity in the force field. In order to maintain smoothness in the formulation of field equations, we assume that the density function $\varrho_{\varkappa}$ is smooth on $\boldsymbol{\varkappa}(\mathscr{B})$.

As in analytical mechanics, the mass measure μ is required to be invariant in any motion of $\mathscr{B}$. Consequently, $\mu(\mathscr{P})$ may be represented also by

$$\mu(\mathscr{P}) = \int_{\boldsymbol{\chi}_t(\mathscr{P})} \varrho \, dx = \int_{\boldsymbol{\varkappa}(\mathscr{P})} \varrho_{\varkappa} \, dX, \tag{15.2}$$

where $\varrho = \varrho(x^i, t)$ is the density function on $\boldsymbol{\chi}_t(\mathscr{B})$. We may rewrite the invariance condition (15.2) in the form

$$\frac{d}{dt} \int_{\boldsymbol{\chi}_t(\mathscr{P})} \varrho \, dx = 0, \qquad \mathscr{P} \subset \mathscr{B}, \tag{15.3}$$

which is called the *principle of conservation of mass*.

The balance principle (15.3) implies the following important identity for any function $\Phi = \Phi(\mathbf{x}, t)$:

$$\frac{d}{dt} \int_{\boldsymbol{\chi}_t(\mathscr{P})} \varrho \Phi \, dx = \int_{\boldsymbol{\chi}_t(\mathscr{P})} \varrho \dot{\Phi} \, dx, \tag{15.4}$$

where $\dot{\Phi}$ denotes the material derivative of Φ. The proof of this identity is more or less obvious. Indeed, since ϱ is the density function of the time-independent mass measure μ, we can rewrite the integral of $\varrho\Phi$ as an integral with respect to dX,

$$\int_{\boldsymbol{\chi}_t(\mathscr{P})} \varrho \Phi \, dx = \int_{\boldsymbol{\varkappa}(\mathscr{P})} \varrho_{\varkappa} \Phi \, dX, \tag{15.5}$$

where the domain of integration $\boldsymbol{\varkappa}(\mathscr{P})$ is independent of t. Hence we can carry the time derivative through the integral sign and obtain

$$\frac{d}{dt} \int_{\boldsymbol{\varkappa}(\mathscr{P})} \varrho_{\varkappa} \Phi \, dX = \int_{\boldsymbol{\varkappa}(\mathscr{P})} \varrho_{\varkappa} \dot{\Phi} \, dX. \tag{15.6}$$

Now after the differentiation with respect to t is taken at each point $\mathbf{X} \in \boldsymbol{\varkappa}(\mathscr{P})$, we transfer the integration back to an integral over $\boldsymbol{\chi}_t(\mathscr{P})$:

$$\int_{\boldsymbol{\varkappa}(\mathscr{P})} \varrho_{\varkappa} \dot{\Phi} \, dX = \int_{\boldsymbol{\chi}_t(\mathscr{P})} \varrho \dot{\Phi} \, dx. \tag{15.7}$$

Combining (15.5)–(15.7), we obtain (15.4).

It should be noted that all preceding equations in this section are stated not for the whole body manifold $\mathscr{B}$ but for an arbitrary subbody $\mathscr{P}$ of $\mathscr{B}$. Of course, these equations remain valid if we replace $\mathscr{P}$ by $\mathscr{B}$, since $\mathscr{B}$ is trivially a subbody of itself. However, for a deformable body manifold $\mathscr{B}$, it is important to consider an arbitrary subbody $\mathscr{P} \subset \mathscr{B}$, since different subbodies of $\mathscr{B}$ generally may deform independently relative to one another. This point reflects again the distinction remarked in the preceding section about the difference between a rigid body and a deformable body.

Next we define the concept of *linear momentum* $\mathbf{L}$ for a motion of $\mathscr{B}$. As we have just remarked, it is important to define this concept not only for the whole body manifold $\mathscr{B}$ but also for all subbodies of $\mathscr{B}$. Therefore we regard $\mathbf{L}$ as being a vector-valued measure on $\mathscr{B}$ at each time t of the motion, such that for any subbody $\mathscr{P} \subset \mathscr{B}$, $\mathbf{L}(t, \mathscr{P})$ is given by the integral

$$\mathbf{L}(t, \mathscr{P}) = \int_{\mathscr{P}} \mathbf{v} \, d\mu = \int_{\boldsymbol{\chi}_t(\mathscr{P})} \varrho \mathbf{v} \, dx, \tag{15.8}$$

where $\mathbf{v}$ is the velocity field defined in Section 14. Using the identity (15.4), we can calculate the rate of change of $\mathbf{L}$ by

$$\frac{d}{dt} \mathbf{L}(t, \mathscr{P}) = \frac{d}{dt} \int_{\boldsymbol{\chi}_t(\mathscr{P})} \varrho \mathbf{v} \, dx = \int_{\boldsymbol{\chi}_t(\mathscr{P})} \varrho \mathbf{a} \, dx, \tag{15.9}$$

where $\mathbf{a}$ is the acceleration field. Clearly, the formulas (15.8) and (15.9) are consistent with the formulas (2.29) and (2.30) for a rigid body.

We define the concept of *moment of momentum* $\mathbf{H}$ relative to the origin $\mathbf{o}$ in a similar way. Namely, $\mathbf{H}$ is a time-dependent vector-valued measure on $\mathscr{B}$ given by the integral

$$\mathbf{H}(t, \mathscr{P}) = \int_{\mathscr{P}} \mathbf{r} \times \mathbf{v} \, d\mu = \int_{\boldsymbol{\chi}_t(\mathscr{P})} \varrho \mathbf{r} \times \mathbf{v} \, dx, \tag{15.10}$$

where $\mathbf{r} = \mathbf{x} - \mathbf{o}$ denotes the position vector relative to $\mathbf{o}$ as before. Then applying (15.4) to $\Phi = \mathbf{r}\times\mathbf{v}$, we get

$$\frac{d}{dt}\mathbf{H}(t,\mathscr{P}) = \frac{d}{dt}\int_{\chi_t(\mathscr{P})} \mathbf{r}\times\mathbf{v}\,dx = \int_{\chi_t(\mathscr{P})} \mathbf{r}\times\mathbf{a}\,dx. \tag{15.11}$$

The formulas (15.10) and (15.11) are generalizations of the formulas (2.31) and (2.39) for a rigid body.

Similarly, we define the concept of *kinetic energy* E: a time-dependent nonnegative measure on $\mathscr{B}$ given by the integral

$$E(t,\mathscr{P}) = \int_{\mathscr{P}} \tfrac{1}{2}\,\|\,\mathbf{v}\,\|^2\,d\mu = \int_{\chi_t(\mathscr{P})} \tfrac{1}{2}\varrho\,\|\,\mathbf{v}\,\|^2\,dx. \tag{15.12}$$

Then its rate of change is

$$\frac{d}{dt}E(t,\mathscr{P}) = \frac{d}{dt}\int_{\chi_t(\mathscr{P})} \tfrac{1}{2}\varrho\,\|\,\mathbf{v}\,\|^2\,dx = \int_{\chi_t(\mathscr{P})} \varrho\mathbf{v}\cdot\mathbf{a}\,dx. \tag{15.13}$$

The formulas (15.12) and (15.13) are generalizations of the formulas (2.40) and (2.41) for a rigid body.

Next we define the concept of a *force system* acting on a body manifold. Again, it is important to consider this concept for each subbody $\mathscr{P}$ of $\mathscr{B}$. In continuum mechanics it is customary to assume that there are only two types of distributed forces acting on each subbody $\mathscr{P}$.

(i) *Body Force* **B**. We assume that **B** is a time-dependent vector valued measure on $\mathscr{B}$ given by the integral

$$\mathbf{B}(t,\mathscr{P}) = \int_{\mathscr{P}} \mathbf{b}\,d\mu = \int_{\chi_t(\mathscr{P})} \varrho\mathbf{b}\,dx, \tag{15.14}$$

where the integrand $\mathbf{b} = \mathbf{b}(\mathbf{x}, t)$ is assumed to be a smooth vector field on $\chi_t(\mathscr{B})$ at each time t of the motion of $\mathscr{B}$. We call **b** the *body force density*.

(ii) *Contact Force* **C**. We assume that **C** is a time-dependent vector-valued measure on each oriented surface in $\mathscr{B}$. In particular, on the boundary $\partial\mathscr{P}$ of $\mathscr{P}$, $\mathbf{C}(t, \partial\mathscr{P})$ is given by the integral

$$\mathbf{C}(t,\partial\mathscr{P}) = \int_{\partial\chi_t(\mathscr{P})} \mathbf{t}\,d\sigma, \tag{15.15}$$

where $d\sigma$ denotes the Euclidean area element on the surface $\partial\chi_t(\mathscr{P})$.

By convention this surface is oriented in such a way that its outward unit normal is regarded as the positive normal vector. We call the integrand $\mathbf{t}$ in (15.15) the *contact force intensity.*

Note. The contact force intensity generally depends on the oriented surface across which it acts. When two oriented surfaces, say, $\mathscr{U}_1$ and $\mathscr{U}_2$, intersect at a certain point, their corresponding contact force intensities need not be the same at that point $\mathbf{x} \in \boldsymbol{\chi}_t(\mathscr{U}_1 \cap \mathscr{U}_2)$. This condition means that $\mathbf{t}$ is not just a function of $(\mathbf{x}, t)$ like the body force density $\mathbf{b}(\mathbf{x}, t)$. Indeed, we should write $\mathbf{t} = \mathbf{t}(\mathbf{x}, t, \partial\boldsymbol{\chi}_t(\mathscr{P}))$ for the integrand in (15.15). We shall discuss the nature of the function $\mathbf{t}$ in Section 16.

We assume that the body force and the contact force are the only forces acting on any subbody $\mathscr{P}$. Then the resultant force on $\mathscr{P}$ at time t is the sum $\mathbf{B}(t, \mathscr{P}) + \mathbf{C}(t, \partial\mathscr{P})$. The balance of linear momentum can now be stated by the following principle:

Cauchy's First Law. Relative to an inertial frame the rate of change of the linear momentum of any subbody $\mathscr{P}$ of $\mathscr{B}$ is equal to the resultant force acting on $\mathscr{P}$.

From (15.9), (15.14), and (15.15) this principle requires that

$$\frac{d}{dt}\int_{\boldsymbol{\chi}_t(\mathscr{P})} \varrho\mathbf{v}\, dx = \int_{\boldsymbol{\chi}_t(\mathscr{P})} \varrho\mathbf{a}\, dx = \int_{\boldsymbol{\chi}_t(\mathscr{P})} \varrho\mathbf{b}\, dx + \int_{\partial\boldsymbol{\chi}_t(\mathscr{P})} \mathbf{t}\, d\sigma. \tag{15.16}$$

Clearly, this principle is consistent with Euler's first law of motion for a rigid body; cf. (4.4).

We assume also that the moment of the body force and the moment of the contact force are the only moments acting on any subbody $\mathscr{P}$. Then the resultant moment on $\mathscr{P}$ is given by

$$\mathbf{G}(t, \mathscr{P}) = \int_{\boldsymbol{\chi}_t(\mathscr{P})} \varrho\mathbf{r}\times\mathbf{b}\, dx + \int_{\partial\boldsymbol{\chi}_t(\mathscr{P})} \mathbf{r}\times\mathbf{t}\, d\sigma. \tag{15.17}$$

The balance of moment of momentum may be stated by the following principle:

Cauchy's Second Law. Relative to the origin of an inertial frame the rate of change of the moment of momentum of any subbody $\mathscr{P}$ of $\mathscr{B}$ is equal to the resultant moment acting on $\mathscr{P}$.

From (15.11) and (15.17) this principle requires that

$$\frac{d}{dt}\int_{\chi_t(\mathscr{P})} \varrho \mathbf{r}\times\mathbf{v}\,dx = \int_{\chi_t(\mathscr{P})} \varrho\mathbf{r}\times\mathbf{a}\,dx$$

$$= \int_{\chi_t(\mathscr{P})} \varrho\mathbf{r}\times\mathbf{b}\,dx + \int_{\partial\chi_t(\mathscr{P})} \mathbf{r}\times\mathbf{t}\,d\sigma. \tag{15.18}$$

Clearly, this principle is consistent with Euler's second law of motion for a rigid body; cf. (4.6).

So far, we have stated the balance principles which govern the motion of a body manifold $\mathscr{B}$ in general. Unlike their counterparts in analytical mechanics, these principles do not determine the motion. Indeed, there are many kinds of deformable bodies in the physical world—solids, fluids, elastic materials, crystalline materials—which must all satisfy these principles. When the same body force and external boundary contact force are applied to different bodies, the resulting motions are generally not the same. The main reason is because the internal contact forces are not determined by any external sources but are due to the internal reactions of the body manifold toward deformations and motions. These reactions generally depend on the particular materials which make up the bodies. In some sense the contact forces play a role in a body manifold similar to that of the constraint forces in a dynamical system. However, continuum mechanics treats far more possibilities for the contact forces than the simple frictionless normal constraint forces considered in the preceding two chapters.

In the following section we shall first set up a general representation for the contact force intensity. Then we develop the mathematical theory for modeling specific classes of materials through the relations between the contact forces and the motions.

16. Cauchy's Postulate and the Stress Principle

At the end of the preceding section we pointed out the importance of the contact force in the theory of deformable bodies. Unfortunately, the contact force is a rather complicated quantity which cannot be characterized easily. Indeed, by its very definition the contact force intensity $\mathbf{t}$ is not just a function of $(\mathbf{x}, t)$ but depends also on the oriented surface across which it acts. Consequently, to specify the contact force in general, we have to assign a function

$$\mathbf{t} = \mathbf{t}(\mathbf{x}, t, \mathscr{U}), \tag{16.1}$$

where $\mathscr{U}$ is an arbitrary oriented surface in the configuration $\boldsymbol{\chi}_t(\mathscr{B})$ of $\mathscr{B}$, and where $\mathbf{x}$ is an arbitrary point in $\mathscr{U}$. Since the collection of oriented surfaces in $\boldsymbol{\chi}_t(\mathscr{B})$ is a very complicated set, it is not easy to characterize the function $\mathbf{t}$.

This difficulty was overcome in an ingenious way by Cauchy, who recognized a special type of contact forces by means of a simple postulate.

Cauchy's Postulate. Suppose that two oriented surfaces $\mathscr{U}_1$ and $\mathscr{U}_2$ have a common positive unit normal at the point $\mathbf{x}$. Then the contact force intensities on $\mathscr{U}_1$ and $\mathscr{U}_2$ coincide at $\mathbf{x}$.

This postulate means, in effect, that the function $\mathbf{t}(\mathbf{x}, t, \mathscr{U})$ depends on $\mathscr{U}$ through the positive unit normal $\mathbf{n}_{\mathscr{U}}(\mathbf{x})$ of $\mathscr{U}$ at $\mathbf{x}$ only. In other words there is a function $\hat{\mathbf{t}}(\mathbf{x}, t, \mathbf{n})$, which is defined for all unit vectors $\mathbf{n} \in \mathscr{V}$ at each $(\mathbf{x}, t)$, such that the function $\mathbf{t}(\mathbf{x}, t, \mathscr{U})$ may be represented by

$$\mathbf{t}(\mathbf{x}, t, \mathscr{U}) = \hat{\mathbf{t}}(\mathbf{x}, t, \mathbf{n}_{\mathscr{U}}(\mathbf{x})). \tag{16.2}$$

Thus the task of characterizing a function of $\mathscr{U}$ is resolved by using a function of a unit vector.

Of course Cauchy's postulate, like the assumption that the constraint surface be frictionless in Lagrangian mechanics, is not a law of physics; it is merely a convenient mathematical condition from which a mathematical model for certain physical objects may be developed. It turns out, however, that the special type of contact forces identified by Cauchy's postulate represents the physical nature of deformable bodies very well. For this reason classical continuum mechanics has dealt almost exclusively with bodies which obey this postulate. Mathematically, it is difficult to find other ways to characterize explicitly the function $\mathbf{t}(\mathbf{x}, t, \mathscr{U})$.

One surprisingly simple and elegant result which follows from Cauchy's postulate is the following theorem:

Cauchy's Stress Principle. Suppose that Cauchy's postulate holds, and suppose that the field $\hat{\mathbf{t}}(\cdot, t, \mathbf{n})$ is continuous on $\boldsymbol{\chi}_t(\mathscr{B})$. Then at each $(\mathbf{x}, t)$ the function $\hat{\mathbf{t}}(\mathbf{x}, t, \cdot)$ is the restriction of a linear function on $\mathscr{V}$ to unit vectors $\mathbf{n}$.

This principle asserts that under the assumption of continuity for the field $\hat{\mathbf{t}}(\cdot, t, \mathbf{n})$, there is a tensor field

$$\mathbf{T} = \mathbf{T}(\mathbf{x}, t): \mathscr{V} \to \mathscr{V}, \qquad \mathbf{x} \in \boldsymbol{\chi}_t(\mathscr{B}), \quad t \in \mathscr{R}, \tag{16.3}$$

such that

$$\hat{\mathbf{t}}(\mathbf{x}, t, \mathbf{n}) = [\mathbf{T}(\mathbf{x}, t)](\mathbf{n}) \tag{16.4}$$

for all unit vectors $\mathbf{n}$ at each $(\mathbf{x}, t)$. Since the set of unit vectors is a generating set for $\mathscr{V}$, the tensor field $\mathbf{T}$, if it exists, is necessarily unique. We call that tensor field $\mathbf{T}$ the field of *stress tensors* on the configuration $\boldsymbol{\chi}_t(\mathscr{B})$.

The existence of the stress tensor may at each $(\mathbf{x}, t)$ be proved by using the original "tetrahedron argument" of Cauchy. We consider the balance of linear momentum (15.16) for the subbody $\mathscr{P}$ such that $\boldsymbol{\chi}_t(\mathscr{P})$ is a tetrahedron with vertices at (x^1, x^2, x^3), $(x^1 + \varepsilon/n^1, x^2, x^3)$, $(x^1, x^2 + \varepsilon/n^2, x^3)$, $(x^1, x^2, x^3 + \varepsilon/n^3)$, where (n^1, n^2, n^3) are the components of a fixed unit vector $\mathbf{n}$, and where ε is a small parameter. We claim that (16.4) follows from the limit of (15.16) as $\varepsilon \to 0$.

Indeed, by using the mean value theorem of integral calculus, we can evaluate (15.16) on the tetrahedron by

$$\begin{aligned}\frac{\varepsilon^3}{6n^1n^2n^3}[\varrho(\mathbf{a} - \mathbf{b})](\hat{\mathbf{x}}, t) &= \frac{\varepsilon^2}{2n^2n^3}\hat{\mathbf{t}}(\mathbf{x}_1, t, -\mathbf{e}_1) + \frac{\varepsilon^2}{2n^1n^3}\hat{\mathbf{t}}(\mathbf{x}_2, t, -\mathbf{e}_2)\\ &\quad + \frac{\varepsilon^2}{2n^1n^2}\hat{\mathbf{t}}(\mathbf{x}_3, t, -\mathbf{e}_3) + \frac{\varepsilon^2}{2n^1n^2n^3}\hat{\mathbf{t}}(\tilde{\mathbf{x}}, t, \mathbf{n}),\end{aligned} \tag{16.5}$$

where $\hat{\mathbf{x}}$ is a certain point in the tetrahedron, and where $\mathbf{x}_1, \mathbf{x}_2, \mathbf{x}_3$ and $\tilde{\mathbf{x}}$ are certain points on the appropriate boundary triangles of the tetrahedron. Dividing (16.5) by ε^2, and then letting $\varepsilon \to 0$, we obtain

$$\hat{\mathbf{t}}(\mathbf{x}, t, \mathbf{n}) = -\hat{\mathbf{t}}(\mathbf{x}, t, -\mathbf{e}_i)n^i, \tag{16.6}$$

where we have used the continuity condition on $\hat{\mathbf{t}}(\cdot, t, \mathbf{n})$ and the fact that $\mathbf{x}_1, \mathbf{x}_2, \mathbf{x}_3$, and $\tilde{\mathbf{x}}$ all approach $\mathbf{x}$ as $\varepsilon \to 0$. Clearly, (16.6) implies that $\hat{\mathbf{t}}(\mathbf{x}, t, \cdot)$ may be extended into a linear function on $\mathscr{V}$, since the coefficients of n^i on the right-hand side of (16.6) depend only on $(\mathbf{x}, t)$. Thus the stress principle is proved.

Setting $\mathbf{n} = \mathbf{e}_j$ in (16.6), we obtain $\hat{\mathbf{t}}(\mathbf{x}, t, \mathbf{e}_j) = -\hat{\mathbf{t}}(\mathbf{x}, t, -\mathbf{e}_j)$. Hence we can rewrite (16.6) as

$$\hat{\mathbf{t}}(\mathbf{x}, t, \mathbf{n}) = \hat{\mathbf{t}}(\mathbf{x}, t, \mathbf{e}_i)n^i = T_i^{\ j}(\mathbf{x}, t)n^i\mathbf{e}_j, \tag{16.7}$$

where $T_i^{\ j}(\mathbf{x}, t)$ are the components of the vectors $\hat{\mathbf{t}}(\mathbf{x}, t, \mathbf{e}_i)$, viz.,

$$\hat{\mathbf{t}}(\mathbf{x}, t, \mathbf{e}_i) = T_i^{\ j}(\mathbf{x}, t)\mathbf{e}_j, \qquad i = 1, 2, 3. \tag{16.8}$$

The component form (16.7) for $\hat{\mathbf{t}}(\mathbf{x}, t, \mathbf{n})$ implies that $T_i^{\ j}(\mathbf{x}, t)$ are also the components of the stress tensor $\mathbf{T}(\mathbf{x}, t)$, viz.,

$$\mathbf{T}(\mathbf{x}, t) = T_i^{\ j}(\mathbf{x}, t)\mathbf{e}_j \otimes \mathbf{e}^i. \tag{16.9}$$

Note. The term "tensor" is derived originally from the word "tension" by virtue of the stress principle. Also, the classical tensor transformation law

$$\bar{T}_i^{\ j} = e_k^{\ j}\bar{e}_i^{\ l}T_l^{\ k} \tag{16.10}$$

may be obtained by changing the basis $\{\mathbf{e}_i\}$ in (16.8) into the basis $\{\bar{\mathbf{e}}_i\} = \{\bar{e}_i^{\ k}\mathbf{e}_k\}$, viz.,

$$\hat{\mathbf{t}}(\mathbf{x}, t, \bar{\mathbf{e}}_i) = \hat{\mathbf{t}}(\mathbf{x}, t, \bar{e}_i^{\ k}\mathbf{e}_k) = \bar{e}_i^{\ k}\hat{\mathbf{t}}(\mathbf{x}, t, \mathbf{e}_k) = \bar{e}_i^{\ k}T_k^{\ l}\mathbf{e}_l = \bar{T}_i^{\ j}\bar{\mathbf{e}}_j = \bar{T}_i^{\ j}\bar{e}_j^{\ l}\mathbf{e}_l, \tag{16.11}$$

which implies (16.10) directly.

The stress principle enables us to represent the contact forces in a simple way by means of the stress tensor field $\mathbf{T}$. Using that tensor field, we can express the balance principles by certain field equations, and we can characterize the material of the body manifold by certain constitutive equations. We shall discuss these items in detail in the following two sections.

17. Field Equations

In Section 15 we formulated three general balance principles which govern the motions of all body manifolds. These principles are stated in integral forms by (15.3), (15.16), and (15.18). In this section we shall derive the differential forms of the balance principles. These forms are called the *field equations*.

First, we introduce an important identity

$$\frac{d}{dt}\int_{\boldsymbol{\chi}_t(\mathscr{P})} \Phi \, dx = \int_{\boldsymbol{\chi}_t(\mathscr{P})} (\dot{\Phi} + \Phi \operatorname{div} \mathbf{v}) \, dx, \tag{17.1}$$

which holds for any smooth function $\Phi = \Phi(\mathbf{x}, t)$. This identity is valid because the time derivative on the left-hand side may be calculated by Leibnitz's rule:

$$\frac{d}{dt}\int_{\boldsymbol{\chi}_t(\mathscr{P})} \Phi \, dx = \int_{\boldsymbol{\chi}_t(\mathscr{P})} \frac{\partial \Phi}{\partial t} \, dx + \int_{\partial\boldsymbol{\chi}_t(\mathscr{P})} \Phi \mathbf{v} \cdot \mathbf{n} \, d\sigma, \tag{17.2}$$

where $\mathbf{n}$ denotes the outward unit normal on $\partial\boldsymbol{\chi}_t(\mathscr{P})$. Notice that the first term on the right-hand side of (17.2) is the rate of change of the integral of Φ when the domain of integration $\boldsymbol{\chi}_t(\mathscr{P})$ is held fixed, and the second term is the flux of Φ through the boundary $\partial\boldsymbol{\chi}_t(\mathscr{P})$ of the domain of integration $\boldsymbol{\chi}_t(\mathscr{P})$. Using the divergence theorem [cf. equation (71.30) in Section 71, IVT-2], we can rewrite the flux term as a volume integral,

$$\int_{\partial\boldsymbol{\chi}_t(\mathscr{P})} \Phi\mathbf{v} \cdot \mathbf{n}\, d\sigma = \int_{\boldsymbol{\chi}_t(\mathscr{P})} \operatorname{div}(\Phi\mathbf{v})\, dx. \tag{17.3}$$

Thus the right-hand side of (17.2) may be reduced to

$$\int_{\boldsymbol{\chi}_t(\mathscr{P})} \left[\frac{\partial\Phi}{\partial t} + \operatorname{div}(\Phi\mathbf{v})\right] dx,$$

which is the same as the right-hand side of (17.1) by virtue of Euler's formula (14.19) for the material derivative $\dot{\Phi}$.

Now applying the identity (17.1) to the balance principle (15.3), we get

$$\frac{d}{dt}\int_{\boldsymbol{\chi}_t(\mathscr{P})} \varrho\, dx = \int_{\boldsymbol{\chi}_t(\mathscr{P})} (\dot{\varrho} + \varrho \operatorname{div} \mathbf{v})\, dx = 0. \tag{17.4}$$

This equation holds for an arbitrary subbody $\mathscr{P}$ of $\mathscr{B}$. Hence if the integrand is continuous, it must vanish identically on the domain $\boldsymbol{\chi}_t(\mathscr{B})$. Thus we obtain the field equation for the conservation of mass:

$$\dot{\varrho} + \varrho \operatorname{div} \mathbf{v} = 0, \tag{17.5}$$

which is also known as the *continuity equation.*

Note. We can derive the identity (17.4) from the identity (17.1) and the continuity equation (17.5), since we have

$$\frac{d}{dt}(\varrho\Phi) + \varrho\Phi \operatorname{div} \mathbf{v} = \varrho\dot{\Phi} + \Phi(\dot{\varrho} + \varrho \operatorname{div} \mathbf{v}) = \varrho\dot{\Phi}. \tag{17.6}$$

Also, the condition (15.2) may be regarded as a change of integration variables from (x^i) to (X^A). Hence the integrands ϱ and ϱ_κ are related by the transformation law:

$$\varrho_\kappa = \varrho \frac{\partial(x^1, x^2, x^3)}{\partial(X^1, X^2, X^3)} = \varrho \det\left[\frac{\partial x^i}{\partial X^A}\right] = \varrho \det \mathbf{F}. \tag{17.7}$$

Taking the material derivative of this equation, we get

$$0 = \dot{\varrho} \det \mathbf{F} + \varrho \frac{\partial}{\partial t}\left(\frac{\partial x^i}{\partial X^A}\right)\frac{\partial X^A}{\partial x^i} \det \mathbf{F} = \det \mathbf{F}(\dot{\varrho} + \varrho \operatorname{div} \mathbf{v}), \tag{17.8}$$

which is equivalent to the continuity equation (17.5).

Next we consider the balance of linear momentum. Using the stress principle (16.4), we can rewrite (15.16) as

$$\int_{\chi_t(\mathscr{P})} \varrho \mathbf{a}\, dx = \int_{\chi_t(\mathscr{P})} \varrho \mathbf{b}\, dx + \int_{\partial\chi_t(\mathscr{P})} \mathbf{T}\mathbf{n}\, d\sigma. \tag{17.9}$$

Since the integrand of the surface integral is a linear function of the outward unit normal $\mathbf{n}$, the divergence theorem may be applied, and we obtain

$$\int_{\chi_t(\mathscr{P})} (\varrho \mathbf{a} - \varrho \mathbf{b} - \operatorname{div} \mathbf{T})\, dx = \mathbf{0}, \tag{17.10}$$

where div $\mathbf{T}$ denotes the vector field with component form

$$\operatorname{div} \mathbf{T} = \frac{\partial T_i^{\,j}}{\partial x^i} \mathbf{e}_j. \tag{17.11}$$

Notice that the coordinate system (x^i) is a rectangular Cartesian system on the physical space $\mathscr{S}$. Relative to such a special coordinate system the contravariant components and the covariant components of any tensor field are numerically the same, and the Christoffel symbols all vanish. The component form (17.11) is valid in such a coordinate system only.

Now since the balance equation (17.10) must be satisfied by an arbitrary subbody $\mathscr{P}$ of $\mathscr{B}$, by the argument as before we obtain the field equation for the balance of linear momentum:

$$\operatorname{div} \mathbf{T} + \varrho \mathbf{b} = \varrho \mathbf{a}, \tag{17.12}$$

which is also known as *Cauchy's equation*.

Like the continuity equation (17.5), Cauchy's equation (17.12) is defined on the instantaneous configuration $\chi_t(\mathscr{B})$, which is a time-dependent domain in the physical space $\mathscr{S}$. For some problems in continuum mechanics it is more convenient to work with fields on the fixed domain $\varkappa(\mathscr{B})$ occupied by $\mathscr{B}$ in the reference configuration. For instance, it is easier to formulate a boundary-value problem when the boundary is fixed.

We proceed now to rewrite the field equation (17.12) as an equation on $\varkappa(\mathscr{B})$.

First, we can rewrite the balance principle (15.16) as

$$\int_{\varkappa(\mathscr{P})} \varrho_\varkappa \mathbf{a}\, dX = \int_{\varkappa(\mathscr{P})} \varrho_\varkappa \mathbf{b}\, dX + \int_{\partial\varkappa(\mathscr{P})} \mathbf{T}_\varkappa \mathbf{N}\, d\Sigma, \tag{17.13}$$

where $\varrho_\varkappa$ denotes the density in $\varkappa(\mathscr{B})$ as before, and where $\mathbf{N}$ and $d\Sigma$ denote the outward unit normal and the Euclidean element of area on the boundary $\partial\varkappa(\mathscr{P})$, respectively. The tensor $\mathbf{T}_\varkappa$ in the integrand of the surface integral is known as the *Piola–Kirchhoff stress tensor.* It is defined by the condition

$$\int_{\varkappa(\mathscr{U})} \mathbf{T}_\varkappa \mathbf{N}\, d\Sigma = \int_{\chi_t(\mathscr{U})} \mathbf{T}\mathbf{n}\, d\sigma \tag{17.14}$$

for any oriented surface $\mathscr{U}$ in $\mathscr{B}$. We can determine the relation between $\mathbf{T}_\varkappa$ and $\mathbf{T}$ in the following way:

Let $(q^\alpha, \alpha = 1, 2)$ be a positive coordinate system on $\mathscr{U}$. Then as usual we can characterize the surfaces $\varkappa(\mathscr{U})$ and $\chi_t(\mathscr{U})$ by

$$\mathbf{X} = \mathbf{X}(q^1, q^2), \qquad \mathbf{x} = \mathbf{x}(q^1, q^2) = \boldsymbol{\varphi}_t(\mathbf{X}(q^1, q^2)). \tag{17.15}$$

From (17.15) we can express the Euclidean elements of area $d\Sigma$ and $d\sigma$ by

$$d\Sigma = \| \mathbf{H}_1 \times \mathbf{H}_2 \|\, dq^1\, dq^2, \qquad d\sigma = \| \mathbf{h}_1 \times \mathbf{h}_2 \|\, dq^1\, dq^2, \tag{17.16}$$

and the positive unit normals $\mathbf{N}$ and $\mathbf{n}$ by

$$\mathbf{N} = \frac{\mathbf{H}_1 \times \mathbf{H}_2}{\| \mathbf{H}_1 \times \mathbf{H}_2 \|}, \qquad \mathbf{n} = \frac{\mathbf{h}_1 \times \mathbf{h}_2}{\| \mathbf{h}_1 \times \mathbf{h}_2 \|}, \tag{17.17}$$

where $\{\mathbf{H}_\alpha\}$ and $\{\mathbf{h}_\alpha\}$ denote the natural bases of (q^α) in $\varkappa(\mathscr{U})$ and $\chi_t(\mathscr{U})$, respectively; i.e.,

$$\mathbf{H}_\alpha = \frac{\partial \mathbf{X}}{\partial q^\alpha}, \qquad \mathbf{h}_\alpha = \frac{\partial \mathbf{x}}{\partial q^\alpha}, \qquad \alpha = 1, 2. \tag{17.18}$$

Now since (q^α) is a convected system, the bases $\{\mathbf{H}_\alpha\}$ and $\{\mathbf{h}_\alpha\}$ are related by

$$\mathbf{h}_\alpha = \mathbf{F}\mathbf{H}_\alpha, \qquad \alpha = 1, 2. \tag{17.19}$$

Substituting (17.16)–(17.19) into (17.14), we obtain

$$\mathbf{T}_\varkappa(\mathbf{H}_1 \times \mathbf{H}_2) = \mathbf{T}(\mathbf{h}_1 \times \mathbf{h}_2) = \mathbf{T}(\mathbf{F}\mathbf{H}_1 \times \mathbf{F}\mathbf{H}_2), \tag{17.20}$$

where $\mathbf{H}_1$ and $\mathbf{H}_2$ may be an arbitrary pair of linearly independent vectors, since the oriented surface $\mathscr{U}$ is arbitrary. We claim that the identity (17.20) holds if and only if $\mathbf{T}_\varkappa$ is related to $\mathbf{T}$ by

$$\mathbf{T}_\varkappa = (\det \mathbf{F})\mathbf{T}(\mathbf{F}^{-1})^T. \tag{17.21}$$

To prove this result, we take the inner product of (17.20) with an arbitrary vector $\mathbf{H}$, obtaining

$$\begin{aligned}\mathbf{H} \cdot \mathbf{T}_\varkappa(\mathbf{H}_1 \times \mathbf{H}_2) &= \mathbf{H} \cdot \mathbf{T}(\mathbf{F}\mathbf{H}_1 \times \mathbf{F}\mathbf{H}_2) = \mathbf{T}^T\mathbf{H} \cdot \mathbf{F}\mathbf{H}_1 \times \mathbf{F}\mathbf{H}_2 \\ &= \mathbf{F}\mathbf{H}_3 \cdot \mathbf{F}\mathbf{H}_1 \times \mathbf{F}\mathbf{H}_2,\end{aligned} \tag{17.22}$$

where $\mathbf{H}_3$ is defined by

$$\mathbf{H}_3 \equiv \mathbf{F}^{-1}\mathbf{T}^T\mathbf{H}. \tag{17.23}$$

Now the triple scalar product $\mathbf{F}\mathbf{H}_3 \cdot \mathbf{F}\mathbf{H}_1 \times \mathbf{F}\mathbf{H}_2$ on the right-hand side of (17.22) is related to the triple scalar product $\mathbf{H}_3 \cdot \mathbf{H}_1 \times \mathbf{H}_2$ by

$$\mathbf{F}\mathbf{H}_3 \cdot \mathbf{F}\mathbf{H}_1 \times \mathbf{F}\mathbf{H}_2 = (\det \mathbf{F})\mathbf{H}_3 \cdot \mathbf{H}_1 \times \mathbf{H}_2; \tag{17.24}$$

cf. (41.27) in Section 41, IVT-1. Substituting (17.24) into (17.22) and observing the fact that $\mathbf{H}_1$, $\mathbf{H}_2$ are arbitrary, we get

$$\mathbf{T}_\varkappa{}^T\mathbf{H} = (\det \mathbf{F})\mathbf{H}_3 = (\det \mathbf{F})(\mathbf{F}^{-1})\mathbf{T}^T\mathbf{H}. \tag{17.25}$$

Now since $\mathbf{H}$ is also arbitrary, (17.25) implies that

$$\mathbf{T}_\varkappa{}^T = (\det \mathbf{F})(\mathbf{F}^{-1})\mathbf{T}^T, \tag{17.26}$$

which is just (17.21) under the operation of transpose.

Having identified the integrands of the balance principle (17.13), we can apply the divergence theorem to the surface integral and obtain

$$\int_{\varkappa(\mathscr{P})} (\varrho_\varkappa \mathbf{a} - \varrho_\varkappa \mathbf{b} - \operatorname{Div} \mathbf{T}_\varkappa)\, dX = \mathbf{0}, \tag{17.27}$$

where $\operatorname{Div} \mathbf{T}_\varkappa$ denotes the vector field:

$$\operatorname{Div} \mathbf{T}_\varkappa = \frac{\partial (T_\varkappa)_A{}^j}{\partial X^A}\, \mathbf{e}_j. \tag{17.28}$$

As before the integral condition (17.27) for an arbitrary subbody $\mathscr{P}$ implies the field equation

$$\operatorname{Div} \mathbf{T}_\varkappa + \varrho_\varkappa \mathbf{b} = \varrho_\varkappa \mathbf{a}, \tag{17.29}$$

which is now defined on the fixed domain $\boldsymbol{\varkappa}(\mathscr{B})$.

Next, we consider the balance of moment of momentum. As before we apply the stress principle (16.4) to the integral form (15.18) of the balance principle, obtaining

$$\int_{\chi_t(\mathscr{P})} \varrho \mathbf{r} \times \mathbf{a}\, dx = \int_{\chi_t(\mathscr{P})} \varrho \mathbf{r} \times \mathbf{b}\, dx + \int_{\partial\chi_t(\mathscr{P})} \mathbf{r} \times \mathbf{T}\mathbf{n}\, d\sigma. \tag{17.30}$$

Again we can rewrite the surface integral as a volume integral by using the divergence theorem. In component form the result is

$$\begin{aligned}\int_{\partial\chi_t(\mathscr{P})} \varepsilon^{ijk} x^j T^{kl} n^l\, d\sigma &= \int_{\chi_t(\mathscr{P})} \varepsilon^{ijk} \frac{\partial}{\partial x^l} (x^j T^{kl})\, dx \\ &= \int_{\chi_t(\mathscr{P})} \varepsilon^{ijk} x^j \frac{\partial T^{kl}}{\partial x^l}\, dx + \int_{\chi_t(\mathscr{P})} \varepsilon^{ijk} T^{kj}\, dx. \end{aligned} \tag{17.31}$$

Substituting this result into (17.30) and eliminating the cross product terms by means of the linear momentum equation (17.12), we get

$$\int_{\chi_t(\mathscr{P})} \varepsilon^{ijk} T^{kj}\, dx = 0. \tag{17.32}$$

Since the subbody $\mathscr{P}$ is arbitrary, this condition implies that

$$\varepsilon^{ijk} T^{kj} = 0, \tag{17.33}$$

which means that $\mathbf{T}$ is a symmetric tensor at each point $\mathbf{x} \in \chi_t(\mathscr{B})$, viz.,

$$\mathbf{T} = \mathbf{T}^T. \tag{17.34}$$

We can express this condition in terms of the Piola–Kirchhoff stress tensor on the fixed domain $\boldsymbol{\varkappa}(\mathscr{B})$ also. Indeed, from (17.21) the condition (17.34) holds if and only if $\mathbf{T}_\varkappa$ satisfies the condition

$$\mathbf{T}_\varkappa \mathbf{F}^T = \mathbf{F} \mathbf{T}_\varkappa{}^T. \tag{17.35}$$

Consequently, $\mathbf{T}_\varkappa$ is generally not a symmetric tensor. Like the condition (17.34), the condition (17.35) characterizes the balance of moment of momentum under the hypothesis that linear momentum is balanced.

18. Constitutive Equations

In Section 15 we remarked that the contact forces in a deformable body are generally caused by the internal reactions of the body toward deformations and motions. In the physical world deformable bodies are made up of many kinds of materials. In the mathematical framework of continuum mechanics each material is represented by a special functional relation between motions and stress tensors. Such a functional relation is called a *constitutive equation*.

We regard a constitutive equation as being a mathematical model which characterizes the mechanical response of a particular body manifold. Using the constitutive equation and the balance equations, we can formulate a deterministic theory for the motions of a body. Of course, the validity of this theory can only be verified by experiments. From the standpoint of mathematics many classes of constitutive equations may be formulated. In application, however, only those classes which give rise to results consistent with experiments deserve our attention.

Note. Physically, a deformable material body possesses many characteristic features, such as color, chemical compositions, heat capacity, electric conductivity, etc. The mechanical response, which is represented mathematically by a constitutive equation, is but one special property of the body. However, all other features are totally irrelevant in the context of continuum mechanics, which is concerned with the mathematical determination of motions of bodies under prescribed external forces only. In particular, if two material bodies with different chemical compositions happen to have exactly the same mechanical response, then in continuum mechanics they can be represented by the same constitutive equation. On the other hand, water and ice are made up of the same chemical substance, H_2O, but since their mechanical responses toward deformations are entirely different, we must model them by different constitutive equations.

In this section we consider only a general class of constitutive equations which was introduced by Noll.[1] He calls the materials represented by these equations *simple materials*. Specifically, a simple material is defined by a

[1] W. Noll, A mathematical theory of the mechanical behavior of continuous media, *Archive for Rational Mechanics and Analysis*, Vol. 2, pp. 197–226, 1958, reprinted in *Continuum Mechanics*, International Science Review Series, Vol. 8, Edited by C. Truesdell, Gordon and Breach, New York, 1965.

constitutive equation of the form

$$\mathbf{T}(\mathbf{x}, t) = \mathbf{G}(\mathbf{F}(\mathbf{X}, t - s), s \in [0, \infty), X, \boldsymbol{\varkappa}), \tag{18.1}$$

which means that the stress tensor $\mathbf{T}(\mathbf{x}, t)$ is determined by the history $\mathbf{F}(\mathbf{X}, t - s)$, $s \in [0, \infty)$, of the deformation gradient of X up to the time t. We call $\mathbf{G}$ the *response functional* of the simple material.

Notice that the response functional generally depends also on the body point X and on the reference configuration $\boldsymbol{\varkappa}$; we shall consider such a dependence in detail later. For simplicity of writing we suppress the notations $\mathbf{x}$, t, $\mathbf{X}$, $s \in [0, \infty)$, X, and $\boldsymbol{\varkappa}$ from the equation (18.1). Thus we write

$$\mathbf{T} = \mathbf{G}(\mathbf{F}(t - s)). \tag{18.2}$$

We call the particular time t in this equation the present time. Then s is the time lapse from the past time $t - s$ to the present time t. The fact that $\mathbf{G}$ may depend on the past values of $\mathbf{F}$ indicates that in general a simple material may have *memory effects*. For the special case in which $\mathbf{G}$ does not depend on the past values of $\mathbf{F}$, however, the material has no memory effects and is called *elastic*.

Noll's general theory of simple materials is based on two basic assertions of invariance.

Principle of Material Frame-Indifference. The response functional $\mathbf{G}$ must satisfy the condition

$$\mathbf{G}(\mathbf{Q}(s)\mathbf{F}(t - s)) = \mathbf{Q}(0)\mathbf{G}(\mathbf{F}(t - s))\mathbf{Q}(0)^T \tag{18.3}$$

for any history of rotations $\mathbf{Q}(s) \in \mathscr{SO}(\mathscr{V})$, $s \in [0, \infty)$.

The meaning of this condition may be explained in the following way: Consider a motion $\bar{\boldsymbol{\chi}}_t: \mathscr{B} \to \mathscr{S}$, $t \in \mathscr{R}$, which differs from the motion $\boldsymbol{\chi}_t: \mathscr{B} \to \mathscr{S}$, $t \in \mathscr{R}$, by a rigid motion only. Then the deformation histories $\bar{\mathbf{F}}(t - s)$ and $\mathbf{F}(t - s)$ of $\bar{\boldsymbol{\chi}}_t$ and $\boldsymbol{\chi}_t$, respectively, are related by

$$\bar{\mathbf{F}}(t - s) = \mathbf{Q}(s)\mathbf{F}(t - s), \tag{18.4}$$

where $\mathbf{Q}(s) \in \mathscr{SO}(\mathscr{V})$ for all $s \in [0, \infty)$. Now from (18.2) the stress tensor $\bar{\mathbf{T}}$ in the motion $\bar{\boldsymbol{\chi}}_t$ is given by

$$\bar{\mathbf{T}} = \mathbf{G}(\bar{\mathbf{F}}(t - s)) = \mathbf{G}(\mathbf{Q}(s)\mathbf{F}(t - s)). \tag{18.5}$$

Consequently, the condition (18.3) amounts to the requirement

$$\bar{\mathbf{T}} = \mathbf{Q}(0)\mathbf{T}\mathbf{Q}(0)^T. \tag{18.6}$$

This requirement can best be explained by means of the contact forces.

We consider an arbitrary oriented surface $\mathscr{U}$ in $\mathscr{B}$ containing the body point X. Let $\mathbf{n}$ and $\bar{\mathbf{n}}$ be the positive unit normals of $\mathscr{U}$ at the positions $\mathbf{x} = \boldsymbol{\chi}_t(X)$ and $\bar{\mathbf{x}} = \bar{\boldsymbol{\chi}}_t(X)$, respectively. Then according to the assumption about $\boldsymbol{\chi}_t$ and $\bar{\boldsymbol{\chi}}_t$, $\mathbf{n}$ and $\bar{\mathbf{n}}$ are related by

$$\bar{\mathbf{n}} = \mathbf{Q}(0)\mathbf{n}. \tag{18.7}$$

Now from the stress principle (16.4) the contact forces $\mathbf{t}$ and $\bar{\mathbf{t}}$ at $\mathbf{x}$ and $\bar{\mathbf{x}}$ are given by

$$\mathbf{t} = \mathbf{T}\mathbf{n}, \qquad \bar{\mathbf{t}} = \bar{\mathbf{T}}\bar{\mathbf{n}}. \tag{18.8}$$

Substituting the condition (18.6) into (18.8), we see that

$$\bar{\mathbf{t}} = \mathbf{Q}(0)\mathbf{t}, \tag{18.9}$$

which means that the contact force $\bar{\mathbf{t}}$ is merely the contact force $\mathbf{t}$ rotated by $\mathbf{Q}(0)$. Conversely, (18.9) is also sufficient for (18.6). Indeed, (18.7)–(18.9) imply that

$$(\bar{\mathbf{T}} - \mathbf{Q}(0)\mathbf{T}\mathbf{Q}(0)^T)\mathbf{n} = \mathbf{0}. \tag{18.10}$$

But $\mathbf{n}$, being the unit normal of an arbitrary surface $\mathscr{U} \subset \mathscr{B}$, is arbitrary; (18.6) follows.

From (18.9) we see that the condition (18.3) reflects our basic understanding that the contact forces are caused by the internal reactions of the body toward deformations and motions. Physically, a rotation $\mathbf{Q}(s)$ from $\boldsymbol{\chi}_{t-s}$ to $\bar{\boldsymbol{\chi}}_{t-s}$ preserves the size, the shape, and the sense of any subbody of $\mathscr{B}$. For this reason we require that the contact forces $\bar{\mathbf{t}}$ and $\mathbf{t}$ be related by the rotation $\mathbf{Q}(0)$ as shown in (18.9).

Note. The condition (18.3) may be viewed also as a transformation law of the constitutive relation (18.1) under a change of frame of reference. Specifically, $\mathbf{Q}(s)\mathbf{F}(t - s)$ and $\mathbf{F}(t - s)$ may be regarded as the deformation histories of the same motion but observed in two different frames of reference. From this point of view the condition (18.9) means simply that the contact forces are frame-indifferent. Because of this explanation[2] the term *material frame-indifference* is chosen for the condition (18.3).

[2] For more details see Chapter 2 in C.-C. Wang and C. Truesdell, *Introduction to Rational Elasticity*, Noordhoff, Leyden, 1973.

It should be pointed out, however, that since $\mathbf{Q}(s)$ is an arbitrary history of rotations, the condition (18.3) involves a change of frame on $\mathscr{E}$ in general, not just a change of inertial frame. In Section 36 of Part B we shall specify the class of (Euclidean) frames in general and the class of inertial frames on $\mathscr{E}$, and we shall discuss changes of frames within these classes in detail. In formulating the classical theories of mechanics, we have decided at the onset to use one particular inertial frame as the frame of reference on $\mathscr{E}$. That special frame enables us to express the equations of motion in the simplest forms. Of course, the classical theories may be formulated in terms of an arbitrary frame on $\mathscr{E}$ or on the basis of $\mathscr{E}$ directly without using any frame of reference at all. Such formulations, however, tend to be far more abstract than ours, which emphasize simple and direct presentations of the subjects.

The general solution of the condition (18.3) is given by the following:

Noll's Representation Theorem. A functional $\mathbf{G}(\mathbf{F}(t-s))$ satisfies the condition (18.3) if and only if it can be represented by

$$\mathbf{G}(\mathbf{F}(t-s)) = \mathbf{R}(t)\mathbf{G}(\mathbf{U}(t-s))\mathbf{R}(t)^T; \tag{18.11}$$

the restriction of $\mathbf{G}$ to positive definite symmetric histories is arbitrary.

We have used the polar decomposition

$$\mathbf{F}(t-s) = \mathbf{R}(t-s)\mathbf{U}(t-s) \tag{18.12}$$

for the deformation gradient $\mathbf{F}$ in (18.11). In view of (18.12) the representation (18.11) means that $\mathbf{G}$ is entirely independent of all past rotations $\mathbf{R}(t-s)$, $s \in (0, \infty)$, and it depends on the present rotation $\mathbf{R}(t)$ in a special way. On the other hand the dependence of $\mathbf{G}$ on the stretch history $\mathbf{U}(t-s)$, $s \in [0, \infty)$, is not restricted at all by the condition of material frame-indifference (18.3).

The proof of (18.11) is more or less obvious. Substituting (18.12) into the left-hand side of (18.11) and then using the condition (18.3), we obtain the right-hand side of (18.11). Thus necessity is proved.

Conversely, if $\mathbf{G}$ is given by (18.11), then we can express $\mathbf{G}(\mathbf{Q}(s)\mathbf{F}(t-s))$ by

$$\begin{aligned}\mathbf{G}(\mathbf{Q}(s)\mathbf{F}(t-s)) &= [\mathbf{Q}(0)\mathbf{R}(t)]\mathbf{G}(\mathbf{U}(t-s))[\mathbf{Q}(0)\mathbf{R}(t)]^T \\ &= \mathbf{Q}(0)[\mathbf{R}(t)\mathbf{G}(\mathbf{U}(t-s))\mathbf{R}(t)^T]\mathbf{Q}(0)^T \\ &= \mathbf{Q}(0)\mathbf{G}(\mathbf{F}(t-s))\mathbf{Q}(0)^T,\end{aligned} \tag{18.13}$$

since the polar decomposition of $\mathbf{Q}(s)\mathbf{F}(t-s)$ is

$$\mathbf{Q}(s)\mathbf{F}(t-s) = [\mathbf{Q}(s)\mathbf{R}(t-s)]\mathbf{U}(t-s). \tag{18.14}$$

Thus (18.11) is also sufficient for (18.3).

Mathematically, the condition of frame-indifference (18.3) corresponds to a transformation law on the functional **G** by a left transformation group. We define a group structure on the collection $\mathscr{Q}$ of rotation histories $\{\mathbf{Q}(s), s \in [0, \infty)\}$ by the binary operation

$$\{\mathbf{Q}(s), s \in [0, \infty)\} * \{\bar{\mathbf{Q}}(s), s \in [0, \infty)\} = \{\mathbf{Q}(s)\bar{\mathbf{Q}}(s), s \in [0, \infty)\}. \tag{18.15}$$

Then we define $\mathscr{Q}$ as a left transformation group on the collection $\mathscr{F}$ of deformation histories $\{\mathbf{F}(t-s), s \in [0, \infty)\}$ by

$$\{\mathbf{Q}(s), s \in [0, \infty)\} \circ \{\mathbf{F}(t-s), s \in [0, \infty)\} = \{\mathbf{Q}(s)\mathbf{F}(t-s), s \in [0, \infty)\}, \tag{18.16}$$

and as a left transformation group on the collection $\mathscr{T}$ of stress tensors by

$$\{\mathbf{Q}(s), s \in [0, \infty)\} \circ \mathbf{T} = \mathbf{Q}(0)\mathbf{T}\mathbf{Q}(0)^T. \tag{18.17}$$

It can be shown easily that the operation $\circ$ satisfies the condition for a left transformation; i.e.,

$$\{\bar{\mathbf{Q}}(s)\} \circ (\{\mathbf{Q}(s)\} \circ \mathbf{T}) = (\{\bar{\mathbf{Q}}(s)\} * \{\mathbf{Q}(s)\}) \circ \mathbf{T} \tag{18.18}$$

for all $\{\bar{\mathbf{Q}}(s)\}$ and $\{\mathbf{Q}(s)\}$ in $\mathscr{Q}$ and for all **T** in $\mathscr{T}$. Using the notations $\mathscr{Q}$, $\mathscr{F}$, and $\mathscr{T}$ just defined, we can regard the response functional **G** as a mapping

$$\mathbf{G}: \mathscr{F} \rightarrow \mathscr{T}. \tag{18.19}$$

Then the condition (18.3) may be regarded as a transformation law of **G** under the action of the group $\mathscr{Q}$:

$$\mathbf{G} \circ \mathscr{Q} = \mathscr{Q} \circ \mathbf{G}. \tag{18.20}$$

That is,

$$\mathbf{G}(\{\mathbf{Q}(s)\} \circ \{\mathbf{F}(t-s)\}) = \{\mathbf{Q}(s)\} \circ \mathbf{G}(\mathbf{F}(t-s)) \tag{18.21}$$

for all $\{\mathbf{Q}(s)\} \in \mathscr{Q}$ and all $\{\mathbf{F}(t-s)\} \in \mathscr{F}$.

Using the transformation group $\mathscr{Q}$, we can define an equivalence relation $\sim$ on the domain $\mathscr{F}$ by

$$\{\mathbf{F}(t-s)\} \sim \{\bar{\mathbf{F}}(t-s)\} \Leftrightarrow \{\bar{\mathbf{F}}(t-s)\} = \{\mathbf{Q}(s)\} \circ \{\mathbf{F}(t-s)\}, \tag{18.22}$$

where $\{\mathbf{Q}(s)\} \in \mathcal{Q}$. We call an equivalence class of this equivalence relation a *coset* or an *orbit* of $\mathcal{Q}$ in $\mathcal{F}$. By virtue of (18.21) the functional $\mathbf{G}$ is determined on each coset by its value at any one representative point in the coset. On the other hand the condition (18.21) does not restrict the values of $\mathbf{G}$ on different cosets at all. This result explains clearly the meaning of Noll's representation formula (18.11). Indeed, the polar decomposition (18.12) asserts that there is one and only one positive definite symmetric history $\{\mathbf{U}(t-s)\}$ in each coset of $\mathcal{Q}$ in $\mathcal{F}$. Using $\{\mathbf{U}(t-s)\}$ as the representative point in the coset, we obtain (18.11) from (18.21). Since (18.21) does not restrict $\mathbf{G}$ on different cosets, the functional $\mathbf{G}(\mathbf{U}(t-s))$ is arbitrary.

Note. The transformation of $\mathcal{F}$ by $\mathcal{Q}$ is called *effective*, since the identity element of $\mathcal{Q}$ is the only element which leaves any one element $\{\mathbf{F}(t-s)\}$ invariant in $\mathcal{F}$. Equivalently, this condition means that the element $\{\mathbf{Q}(s)\}$ in (18.22) is unique. Because $\mathcal{Q}$ is effective on $\mathcal{F}$, the value of $\mathbf{G}$ at any representative point in a coset is entirely arbitrary in the set $\mathcal{T}$. Otherwise, the value of $\mathbf{G}$ at any point $\{\mathbf{F}(t-s)\}$ must be invariant under all $\{\mathbf{Q}(s)\}$ which preserve the point $\{\mathbf{F}(t-s)\}$. This necessary condition becomes vacuous, however, when $\mathcal{Q}$ is effective on $\mathcal{F}$, since in this case $\{\mathbf{F}(t-s)\}$ is preserved by the identity element $\{\mathbf{I}(s)\}$ of $\mathcal{Q}$ only, and $\{\mathbf{I}(s)\}$ clearly preserves each element $\mathbf{T}$ in $\mathcal{T}$.

Next, we consider another condition of invariance on $\mathbf{G}$ due to the rule of material symmetry. This rule reflects the observation that in the physical world a deformable material body generally possesses certain material symmetry. For example, if the body is a crystal, then two configurations of the body differing by a transformation belonging to the crystallographic group are indistinguishable. In the context of simple materials we can define the physical concept of indistinguishable configurations by the condition that the response functionals relative to the configurations be identical. Specifically, we say that the reference configurations $\boldsymbol{\varkappa}$ and $\bar{\boldsymbol{\varkappa}}$ are *materially isomorphic* at the point $X \in \mathcal{B}$ if

$$\mathbf{G}(\mathbf{F}(\mathbf{X}, t-s), s \in [0, \infty), X, \boldsymbol{\varkappa}) = \mathbf{G}(\mathbf{F}(\bar{\mathbf{X}}, t-s), s \in [0, \infty), X, \bar{\boldsymbol{\varkappa}}) \quad (18.23)$$

whenever $\mathbf{F}(\mathbf{X}, \cdot) = \mathbf{F}(\bar{\mathbf{X}}, \cdot)$. Here $\mathbf{X}$ and $\bar{\mathbf{X}}$ denote the positions occupied by X in the reference configurations $\boldsymbol{\varkappa}$ and $\bar{\boldsymbol{\varkappa}}$, respectively. We can express the condition (18.23) in terms of the response functional $\mathbf{G}(\mathbf{F}(t-s))$ relative to a single reference configuration $\boldsymbol{\varkappa}$ in the following way:

In general, when we change the reference configuration from $\varkappa$ to $\hat{\varkappa}$, the response functional is changed according to the transformation law:

$$\mathbf{G}(\hat{\mathbf{F}}(\hat{\mathbf{X}}, t - s), s \in [0, \infty), X, \hat{\varkappa}) = \mathbf{G}(\hat{\mathbf{F}}(\hat{\mathbf{X}}, t - s)\mathbf{K}(\mathbf{X}), s \in [0, \infty), X, \varkappa), \tag{18.24}$$

where $\mathbf{K}(\mathbf{X})$ denotes the deformation gradient from $\varkappa$ to $\hat{\varkappa}$ at the point X. This transformation law is valid since for the same motion $\boldsymbol{\chi}_t$ the deformation gradients $\mathbf{F}(\mathbf{X}, t - s)$ and $\hat{\mathbf{F}}(\hat{\mathbf{X}}, t - s)$ relative to $\varkappa$ and $\hat{\varkappa}$, respectively, are related by the chain rule

$$\hat{\mathbf{F}}(\hat{\mathbf{X}}, t - s) = \mathbf{F}(\mathbf{X}, t - s)\mathbf{K}(\mathbf{X}). \tag{18.25}$$

Consequently, the stress tensor $\mathbf{T}(\mathbf{x}, t)$ may be obtained either by the left-hand side or by the right-hand side of the equation (18.24).

Now using the general transformation law (18.24), we can rewrite the condition (18.23) as

$$\mathbf{G}(\mathbf{F}(\mathbf{X}, t - s), s \in [0, \infty), X, \varkappa) = \mathbf{G}(\mathbf{F}(\mathbf{X}, t - s)\mathbf{H}, s \in [0, \infty), X, \varkappa), \tag{18.26}$$

where both sides are referred to the same configuration $\varkappa$. We call the deformation gradient $\mathbf{H}$ from $\varkappa$ to $\bar{\varkappa}$ a *material automorphism* of X relative to the reference configuration $\varkappa$. It can be proved easily that the set of all material automorphisms of X relative to $\varkappa$ form a group $\mathscr{G} = \mathscr{G}(X, \varkappa)$. We call this group the *symmetry group* (or the *isotropy group*) of X relative to $\varkappa$. Thus we have the following condition due to the rule of symmetry:

Condition of Material Symmetry. The response functional $\mathbf{G}$ must satisfy the condition

$$\mathbf{G}(\mathbf{F}(t - s)) = \mathbf{G}(\mathbf{F}(t - s)\mathbf{H}) \tag{18.27}$$

for any $\mathbf{H}$ belonging to the symmetry group $\mathscr{G}$.

From our explanation of material symmetry the condition (18.27) appears to be the definition for the symmetry group $\mathscr{G}$ rather than a condition for the response functional $\mathbf{G}$. In application, however, the functional $\mathbf{G}$ is rarely given explicitly, while the group $\mathscr{G}$ can often be assumed. As a result, (18.27) becomes a condition upon the unspecified response functional $\mathbf{G}$. In other words an assignment of the symmetry group narrows the class of response functionals. The conditions of frame-indifference

(18.3) and symmetry (18.27) are the two basic assertions of invariance which have been exploited for a simple material in general.

Mathematically, the condition (18.27) may be viewed as a transformation law to which the mapping $\mathbf{G}$ defined by (18.19) is subject under the action of a right transformation group $\mathbf{G}$. Specifically, we define $\mathbf{G}$ as a right transformation group on $\mathscr{F}$ by

$$\{\mathbf{F}(t-s), s \in [0, \infty)\} \circ \mathbf{H} \equiv \{\mathbf{F}(t-s)\mathbf{H}, s \in [0, \infty)\}, \tag{18.28}$$

and as a right transformation group on $\mathscr{T}$ by

$$\mathbf{T} \circ \mathbf{H} = \mathbf{T} \tag{18.29}$$

for all $\mathbf{H} \in \mathscr{G}$. Using the definitions (18.28) and (18.29), we can rewrite (18.27) as the transformation law

$$\mathbf{G} \circ \mathscr{G} = \mathscr{G} \circ \mathbf{G}. \tag{18.30}$$

That is,

$$\mathbf{G}(\{\mathbf{F}(t-s)\} \circ \mathbf{H}) = \mathbf{G}(\mathbf{F}(t-s)) \circ \mathbf{H} \tag{18.31}$$

for all $\{\mathbf{F}(t-s)\} \in \mathscr{F}$ and all $\mathbf{H} \in \mathscr{G}$.

We define a coset of $\mathscr{G}$ as before: an equivalence class induced by the equivalence relation

$$\{\mathbf{F}(t-s)\} \approx \{\bar{\mathbf{F}}(t-s)\} \Leftrightarrow \{\bar{\mathbf{F}}(t-s)\} = \{\mathbf{F}(t-s)\} \circ \mathbf{H}, \tag{18.32}$$

where $\mathbf{H} \in \mathscr{G}$. Then we can describe a general solution of the condition (18.30) as before: $\mathbf{G}$ satisfies (18.30) if and only if it is constant on each coset of $\mathscr{G}$ in $\mathscr{F}$. On different cosets the constant values of $\mathbf{G}$ are arbitrary.

A general solution for both (18.3) and (18.27) has been obtained by Wang.[3] We shall consider some important special cases of representations in the following section.

19. Some Representation Theorems

In the preceding section we have introduced two basic conditions of invariance on a constitutive equation, namely, the condition of material frame-indifference (18.3) and the condition of material symmetry (18.27). We shall now derive some results from these conditions.

[3] C.-C. Wang, A general representation theorem for constitutive relations, *Archive for Rational Mechanics and Analysis*, Vol. 32, pp. 1–25, 1969.

First, we remark that the symmetry group $\mathscr{G}$ of any simple material is generally contained in the special linear group, viz.,

$$\mathscr{G} \subset \mathscr{SL}(\mathscr{V}). \tag{19.1}$$

This restriction reflects the observation that the stress tensors in configurations of different mass densities are generally different. Consequently, in order to satisfy the condition (18.27), the determinants of $\mathbf{F}(t)\mathbf{H}$ and $\mathbf{F}(t)$ must be equal; i.e.,

$$\det \mathbf{H} = 1. \tag{19.2}$$

From this observation we see that the largest possible symmetry group is $\mathscr{G} = \mathscr{SL}(\mathscr{V})$. Noll calls a simple material having that symmetry group a *simple fluid.*

It should be noted that the symmetry group of a simple material generally depends on the reference configuration $\boldsymbol{\varkappa}$. Under any change of reference configuration, say from $\boldsymbol{\varkappa}$ to $\hat{\boldsymbol{\varkappa}}$, the transformation law for $\mathscr{G}$ is given by Noll:

$$\mathscr{G}(X, \hat{\boldsymbol{\varkappa}}) = \mathbf{K}\mathscr{G}(X, \boldsymbol{\varkappa})\mathbf{K}^{-1}, \tag{19.3}$$

where $\mathbf{K}$ denotes the deformation gradient at X from $\boldsymbol{\varkappa}$ to $\hat{\boldsymbol{\varkappa}}$. This transformation law follows directly from (18.24). Indeed, if $\mathbf{H}$ satisfies

$$\mathbf{G}(\mathbf{F}(t-s)\mathbf{H}, X, \boldsymbol{\varkappa}) = \mathbf{G}(\mathbf{F}(t-s), X, \boldsymbol{\varkappa}) \tag{19.4}$$

for all $\{\mathbf{F}(t-s)\} \in \mathscr{F}$, then from (18.24) we have

$$\begin{aligned} \mathbf{G}(\mathbf{F}(t-s)\mathbf{KHK}^{-1}, X, \hat{\boldsymbol{\varkappa}}) &= \mathbf{G}(\mathbf{F}(t-s)\mathbf{KHK}^{-1}\mathbf{K}, X, \boldsymbol{\varkappa}) \\ &= \mathbf{G}(\mathbf{F}(t-s)\mathbf{KH}, X, \boldsymbol{\varkappa}) = \mathbf{G}(\mathbf{F}(t-s)\mathbf{K}, X, \boldsymbol{\varkappa}) \\ &= \mathbf{G}(\mathbf{F}(t-s), X, \hat{\boldsymbol{\varkappa}}). \end{aligned} \tag{19.5}$$

Thus $\mathscr{G}(X, \hat{\boldsymbol{\varkappa}}) \supset \mathbf{K}\mathscr{G}(X, \boldsymbol{\varkappa})\mathbf{K}^{-1}$. By reversing the roles of $\boldsymbol{\varkappa}$ and $\hat{\boldsymbol{\varkappa}}$, we have also $\mathscr{G}(X, \boldsymbol{\varkappa}) \supset \mathbf{K}^{-1}\mathscr{G}(X, \hat{\boldsymbol{\varkappa}})\mathbf{K}$. From these two inclusions Noll's transformation rule (19.3) follows.

By virtue of the transformation rule (19.3) the symmetry groups of a simple material relative to different reference configurations form a conjugate class of subgroups in the special linear group. For the special case when $\mathscr{G}$ coincides with $\mathscr{SL}(\mathscr{V})$; i.e., for a simple fluid, the conjugate class consists of a single subgroup, namely, $\mathscr{SL}(\mathscr{V})$ itself. Hence in this case the symmetry group is independent of the reference configuration.

Now for a simple fluid the response functional **G** must satisfy the following pair of conditions:

$$\mathbf{G}(\mathbf{F}(t-s)\mathbf{H}) = \mathbf{G}(\mathbf{F}(t-s)), \qquad \mathbf{H} \in \mathscr{SL}(\mathscr{V}), \tag{19.6}$$

$$\mathbf{G}(\mathbf{Q}(s)\mathbf{F}(t-s)) = \mathbf{Q}(0)\mathbf{G}(\mathbf{F}(t-s))\mathbf{Q}(0)^T, \qquad \{\mathbf{Q}(s)\} \in \mathscr{Q}. \tag{19.7}$$

In order to exploit these conditions we first rewrite the response functional **G** as a functional $\tilde{\mathbf{G}}$ of the form

$$\mathbf{G}(\mathbf{F}(t-s)) = \tilde{\mathbf{G}}(\mathbf{F}_t(t-s), \mathbf{F}(t)), \tag{19.8}$$

where $\mathbf{F}_t(t-s)$ denotes the relative deformation history; i.e., the deformation history relative to the present configuration

$$\mathbf{F}_t(t-s) \equiv \mathbf{F}(t-s)\mathbf{F}(t)^{-1}, \qquad s \in [0, \infty). \tag{19.9}$$

In terms of the functional $\tilde{\mathbf{G}}$ the conditions (19.6) and (19.7) become

$$\tilde{\mathbf{G}}(\mathbf{F}_t(t-s), \mathbf{F}(t)\mathbf{H}) = \tilde{\mathbf{G}}(\mathbf{F}_t(t-s), \mathbf{F}(t)), \tag{19.10}$$

$$\tilde{\mathbf{G}}(\mathbf{Q}(s)\mathbf{F}_t(t-s)\mathbf{Q}(0)^T, \mathbf{Q}(0)\mathbf{F}(t)) = \mathbf{Q}(0)\tilde{\mathbf{G}}(\mathbf{F}_t(t-s), \mathbf{F}(t))\mathbf{Q}(0)^T, \tag{19.11}$$

where $\mathbf{H} \in \mathscr{SL}(\mathscr{V})$ and $\{\mathbf{Q}(s)\} \in \mathscr{Q}$ as before. We now show that these conditions may be reduced in the following way:

Theorem (Noll). A functional $\tilde{\mathbf{G}}(\mathbf{F}_t(t-s), \mathbf{F}(t))$ obeys the conditions (19.10) and (19.11) if and only if it can be represented by

$$\tilde{\mathbf{G}}(\mathbf{F}_t(t-s), \mathbf{F}(t)) = \hat{\mathbf{G}}(\mathbf{U}_t(t-s), \varrho(t)), \tag{19.12}$$

where $\hat{\mathbf{G}}$ is an *isotropic functional*; i.e., it satisfies the condition

$$\hat{\mathbf{G}}(\mathbf{Q}\mathbf{U}_t(t-s)\mathbf{Q}^T, \varrho(t)) = \mathbf{Q}\hat{\mathbf{G}}(\mathbf{U}_t(t-s), \varrho(t))\mathbf{Q}^T \tag{19.13}$$

for all orthogonal tensors $\mathbf{Q} \in \mathscr{O}(\mathscr{V})$.

To prove this theorem we remark that a general solution of the condition (19.10) is given by

$$\tilde{\mathbf{G}}(\mathbf{F}_t(t-s)), \mathbf{F}(t) = \hat{\mathbf{G}}(\mathbf{F}_t(t-s), \varrho(t)), \tag{19.14}$$

since first, each coset of $\mathscr{SL}(\mathscr{V})$ may be characterized by the determinant of $\mathbf{F}(t)$, viz.,

$$\bar{\mathbf{F}}(t) = \mathbf{F}(t)\mathbf{H}, \mathbf{H} \in \mathscr{SL}(\mathscr{V}) \Leftrightarrow \det \bar{\mathbf{F}}(t) = \det \mathbf{F}(t), \tag{19.15}$$

and second, by virtue of (17.7) det $\mathbf{F}(t)$ is characterized by the density $\varrho(t)$.

Now after $\tilde{\mathbf{G}}$ is reduced by (19.14), the condition (19.11) may be expressed in terms of $\hat{\mathbf{G}}$ as

$$\hat{\mathbf{G}}(\mathbf{Q}(s)\mathbf{F}_t(t-s)\mathbf{Q}(0)^T, \varrho(t)) = \mathbf{Q}(0)\hat{\mathbf{G}}(\mathbf{F}_t(t-s), \varrho(t))\mathbf{Q}(0)^T. \qquad (19.16)$$

Then choosing $\mathbf{Q}(s) = \mathbf{R}_t(t-s)^T$ and $\mathbf{Q}(0) = \mathbf{I}$, we obtain (19.12). Similarly, choosing $\mathbf{Q}(0) = \mathbf{Q}$ and $\mathbf{Q}(s) = \mathbf{I}$ for all $s \in (0, \infty)$, we obtain (19.13). Thus necessity of (19.12) and (19.13) is proved.

Conversely, when (19.12) and (19.13) are satisfied, we show the validity of (19.16) by

$$\begin{aligned}
&\hat{\mathbf{G}}(\mathbf{Q}(s)\mathbf{F}_t(t-s)\mathbf{Q}(0)^T, \varrho(t))\\
&\quad = \hat{\mathbf{G}}(\mathbf{Q}(s)\mathbf{R}_t(t-s)\mathbf{Q}(0)^T\mathbf{Q}(0)\mathbf{U}_t(t-s)\mathbf{Q}(0)^T, \varrho(t))\\
&\quad = \hat{\mathbf{G}}(\mathbf{Q}(0)\mathbf{U}_t(t-s)\mathbf{Q}(0)^T, \varrho(t)) = \mathbf{Q}(0)\hat{\mathbf{G}}(\mathbf{U}_t(t-s), \varrho(t))\mathbf{Q}(0)^T\\
&\quad = \mathbf{Q}(0)\hat{\mathbf{G}}(\mathbf{F}_t(t-s), \varrho(t))\mathbf{Q}(0)^T. \qquad (19.17)
\end{aligned}$$

Thus sufficiency of (19.12) and (19.13) is proved.

Note. In the preceding proof the tensor $\mathbf{Q}(s)$ is a rotation (i.e., proper orthogonal) at each $s \in [0, \infty)$. The condition (19.13), however, is valid for any improper orthogonal tensor $\mathbf{Q}$ also, since it is quadratic in $\mathbf{Q}$.

Although a general solution of (19.13) is not known, we can often use the condition directly to obtain valuable information about the functional $\hat{\mathbf{G}}$. For example, if the relative right stretch history $\{\mathbf{U}_t(t-s)\}$ is a *plane deformation*, i.e., all $\mathbf{U}_t(t-s)$, $s \in [0, \infty)$, share a common proper vector $\mathbf{f}$, then the stress tensor $\mathbf{T}$ given by the value of $\hat{\mathbf{G}}$ at the history must also have $\mathbf{f}$ as a proper vector. This assertion can be proved by a standard argument for an isotropic function. We choose $\mathbf{Q}$ to be the orthogonal reflection with respect to the vector $\mathbf{f}$. Then $\mathbf{Q}$ commutes with $\mathbf{U}_t(t-s)$; i.e.,

$$\mathbf{Q}\mathbf{U}_t(t-s)\mathbf{Q}^T = \mathbf{U}_t(t-s). \qquad (19.18)$$

We verify this fact by using an orthonormal basis of the form $\{\mathbf{f}_1, \mathbf{f}_2, \mathbf{f}\}$. Relative to this basis the component matrix of $\mathbf{U}_t(t-s)$ is a block matrix with a general 2×2 block in $\{\mathbf{f}_1, \mathbf{f}_2\}$ and a proper number in $\mathbf{f}$, and the component matrix of $\mathbf{Q}$ is a diagonal matrix with diagonal elements $(1, 1, -1)$. Clearly, these two matrices commute.

Now from (19.18) and (19.13) we see that the reflection $\mathbf{Q}$ must also commute with the stress tensor $\mathbf{T}$, which is given by the functional $\hat{\mathbf{G}}$ at the history $\mathbf{U}_t(t-s)$ in accord with the representation (19.12). Then by using the component matrix of $\mathbf{T}$ relative to the basis $\{\mathbf{f}_1, \mathbf{f}_2, \mathbf{f}\}$, we see that $\mathbf{f}$ must be a proper vector of $\mathbf{T}$ also. In other words $\mathbf{T}$ is a plane stress in the same plane of the stretch history $\mathbf{U}_t(t-s)$. Results like this have been used by Coleman and Noll[4] to analyze the viscometric flows of a simple fluid in general; cf. Section 23 in Chapter 4.

The main difficulty in finding a general solution for the condition of isotropy (19.13) is due to the fact that $\hat{\mathbf{G}}$ generally depends on all past values of the right stretch history $\mathbf{U}_t(t-s)$. To simplify this situation, we consider next a simple fluid which has only infinitesimal memory effects. Specifically, we assume that $\hat{\mathbf{G}}$ depends on $\mathbf{U}_t(t-s)$ only through its derivative with respect to s at $s=0$, i.e., the constitutive equation reduces to

$$\mathbf{T} = \hat{\mathbf{G}}(\mathbf{D}, \varrho), \tag{19.19}$$

where $\mathbf{D}$ denotes the stretching tensor defined by (14.28). The condition (19.13) now implies that $\hat{\mathbf{G}}$ is an isotropic function; i.e.,

$$\hat{\mathbf{G}}(\mathbf{QDQ}^T, \varrho) = \mathbf{Q}\hat{\mathbf{G}}(\mathbf{D}, \varrho)\mathbf{Q}^T \tag{19.20}$$

for all $\mathbf{Q} \in \mathscr{O}(\mathscr{V})$. A general solution of this condition is known.

Theorem (Rivlin and Ericksen[5]). A function $\hat{\mathbf{G}}$ obeys the condition (19.20) if and only if it can be represented by

$$\hat{\mathbf{G}}(\mathbf{D}, \varrho) = \gamma_0\mathbf{I} + \gamma_1\mathbf{D} + \gamma_2\mathbf{D}^2, \tag{19.21}$$

where $\gamma_0, \gamma_1, \gamma_2$ are functions of ϱ and the three fundamental invariants of $\mathbf{D}$, viz.,

$$\mathrm{I}_{\mathbf{D}} = \operatorname{tr}\mathbf{D}, \qquad \mathrm{II}_{\mathbf{D}} = \tfrac{1}{2}[(\operatorname{tr}\mathbf{D})^2 - \operatorname{tr}\mathbf{D}^2], \qquad \mathrm{III}_{\mathbf{D}} = \det\mathbf{D}; \tag{19.22}$$

cf. (26.7) in Section 26, IVT-1.

(4) B. D. Coleman and W. Noll, On certain steady flows of general fluids, *Archive for Rational Mechanics and Analysis*, Vol. 3, pp. 289–303, 1959. Reprinted in *Continuum Mechanics*, International Science Review Series, Vol. 8, Edited by C. Truesdell, Gordon and Breach, New York, 1965.

(5) R. S. Rivlin and J. L. Ericksen, Stress-deformation relations for isotropic materials, *Journal of Rational Mechanics and Analysis*, Vol. 4, pp. 323–425, 1955. When restricted to polynomial functions this theorem is a special case of a general result in classical invariant theory. The discovery of this special case in the context of continuum mechanics may be attributed to many authors, including Finger (1894), Reiner and Prager (1945), Reiner and Richter (1948), and others.

The proof of this theorem follows from the remark about isotropic functions mentioned before, namely, a proper vector of **D** is necessarily a proper vector of **T**. From this remark we see that there are three possibilities:

(i) **D** has only one distinct proper number, so that

$$\mathbf{D} = d\mathbf{I}. \tag{19.23}$$

In this case the remark implies that **T** has only one distinct proper number also. Hence

$$\hat{\mathbf{G}}(\mathbf{D}, \varrho) = \gamma_0 \mathbf{I}. \tag{19.24}$$

(ii) **D** has only two distinct proper numbers, so that

$$\mathbf{D} = c\mathbf{I} + d\mathbf{f} \otimes \mathbf{f}, \tag{19.25}$$

where $d \neq 0$. In this case the remark implies that **T** has at most two distinct proper numbers; moreover, $\hat{\mathbf{G}}$ has the form

$$\hat{\mathbf{G}}(\mathbf{D}, \varrho) = \gamma_0 \mathbf{I} + \gamma_1 \mathbf{D}, \tag{19.26}$$

since the tensor space generated by $\{\mathbf{I}, \mathbf{f} \otimes \mathbf{f}\}$ coincides with that generated by $\{\mathbf{I}, \mathbf{D}\}$.

(iii) **D** has three distinct proper numbers, so that

$$\mathbf{D} = d_1 \mathbf{f}_1 \otimes \mathbf{f}_1 + d_2 \mathbf{f}_2 \otimes \mathbf{f}_2 + d_3 \mathbf{f}_3 \otimes \mathbf{f}_3, \tag{19.27}$$

where $\{\mathbf{f}_i\}$ is an orthonormal basis, and where d_1, d_2, d_3 are unequal. Then the remark implies that $\{\mathbf{f}_i\}$ is also a principal basis for **T**; moreover, $\hat{\mathbf{G}}$ has the form

$$\hat{\mathbf{G}}(\mathbf{D}, \varrho) = \gamma_0 \mathbf{I} + \gamma_1 \mathbf{D} + \gamma_2 \mathbf{D}^2, \tag{19.28}$$

since the tensor space generated by $\{\mathbf{f}_1 \otimes \mathbf{f}_1, \mathbf{f}_2 \otimes \mathbf{f}_2, \mathbf{f}_3 \otimes \mathbf{f}_3\}$ coincides with that generated by $\{\mathbf{I}, \mathbf{D}, \mathbf{D}^2\}$.

Now the coefficients $\gamma_0, \gamma_1, \gamma_2$ in (19.24), (19.26), and (19.28) are functions of **D** and ϱ. However, since

$$\mathbf{QIQ}^T = \mathbf{I}, \qquad \mathbf{QD}^2\mathbf{Q}^T = \mathbf{QDQ}^T\mathbf{QDQ}^T = (\mathbf{QDQ}^T)^2, \tag{19.29}$$

the condition (19.20) implies further that $\gamma_0, \gamma_1, \gamma_2$ are isotropic functions of **D**; i.e.,

$$\gamma_a(\mathbf{QDQ}^T, \varrho) = \gamma_a(\mathbf{D}, \varrho), \qquad a = 0, 1, 2, \tag{19.30}$$

for all $\mathbf{Q} \in \mathscr{O}(\mathscr{V})$. As a result, γ_a depend on $\mathbf{D}$ only through its fundamental invariants $\mathrm{I}_{\mathbf{D}}, \mathrm{II}_{\mathbf{D}}, \mathrm{III}_{\mathbf{D}}$, since a symmetric tensor $\bar{\mathbf{D}}$ is related to $\mathbf{D}$ by

$$\bar{\mathbf{D}} = \mathbf{Q}\mathbf{D}\mathbf{Q}^T \tag{19.31}$$

if and only if its proper numbers, or equivalently its fundamental invariants, are the same as those of $\mathbf{D}$. Thus necessity of (19.21) is proved.

Sufficiency of (19.21) is entirely obvious, since from (19.29) and (19.30) we obtain (19.20). Thus the theorem is proved.

The special class of simple fluids defined by the constitutive equation

$$\mathbf{T} = \gamma_0\mathbf{I} + \gamma_1\mathbf{D} + \gamma_2\mathbf{D}^2 \tag{19.32}$$

is called the class of (compressible) *Reiner–Rivlin fluids*. This class includes the classical (compressible) *Newtonian fluids* as special cases. For a Newtonian fluid the response function depends linearly on the stretching tensor $\mathbf{D}$, so its constitutive equation is of the form

$$\mathbf{T} = -p\mathbf{I} + \lambda(\operatorname{tr}\mathbf{D})\mathbf{I} + \mu\mathbf{D}, \tag{19.33}$$

where λ, μ, and p are functions of ϱ.

When a simple fluid has no memory effect at all, i.e., when it is elastic as defined before, its constitutive equation has the form

$$\mathbf{T} = -p\mathbf{I}, \tag{19.34}$$

where the pressure p is a function of the density ϱ. In classical fluid mechanics this material is known as the (compressible) *Eulerian fluid.*

So far, we have considered the condition of material symmetry for the group $\mathscr{G} = \mathscr{SL}(\mathscr{V})$. Since this group is the largest possible symmetry group of a simple material, any solution of the conditions (19.6) and (19.7) is automatically a solution of a similar pair of conditions but for a smaller group $\mathscr{G}$. Noll calls a *simple solid* a simple material for which there is a reference configuration $\varkappa$ relative to which $\mathscr{G}$ is contained in the rotation group, viz.,

$$\mathscr{G} \subset \mathscr{SO}(\mathscr{V}). \tag{19.35}$$

Note. Unlike the symmetry group of a simple fluid, the symmetry group of a simple solid may vary when we change the reference configuration. Specifically, the transformation law is given by Noll's rule (19.3), which does not always preserve the condition (19.35). As a result, the symmetry group of a simple solid may or may not be a subgroup of $\mathscr{SO}(\mathscr{V})$.

The validity of (19.35) depends very much on the choice of the reference configuration. We call $\varkappa$ an *undistorted reference configuration* when the condition (19.35) holds.

For a simple solid in general the largest possible symmetry group (relative to an undistorted reference configuration) is the rotation group $\mathscr{SO}(\mathscr{V})$ itself. A simple solid having this symmetry group is called an *isotropic simple solid*. Its response functional $\mathbf{G}$ satisfies the condition (19.7) and the condition

$$\mathbf{G}(\mathbf{F}(t-s)\mathbf{Q}) = \mathbf{G}(\mathbf{F}(t-s)), \qquad \mathbf{Q} \in \mathscr{SO}(\mathscr{V}). \tag{19.36}$$

As before we can rewrite $\mathbf{G}$ as $\tilde{\mathbf{G}}$ by (19.8). Then the condition (19.36) is equivalent to the condition

$$\tilde{\mathbf{G}}(\mathbf{F}_t(t-s), \mathbf{F}(t)\mathbf{Q}) = \mathbf{G}(\mathbf{F}_t(t-s), \mathbf{F}(t)). \tag{19.37}$$

Using an argument similar to the proof of (19.12) and (19.13), we have the following result:

Theorem (Noll). A functional $\tilde{\mathbf{G}}(\mathbf{F}_t(t-s), \mathbf{F}(t))$ obeys the conditions (19.11) and (19.37) if and only if it can be represented by

$$\tilde{\mathbf{G}}(\mathbf{F}_t(t-s), \mathbf{F}(t)) = \hat{\mathbf{G}}(\mathbf{U}_t(t-s), \mathbf{V}(t)), \tag{19.38}$$

where $\hat{\mathbf{G}}$ is an isotropic functional; i.e.,

$$\hat{\mathbf{G}}(\mathbf{Q}\mathbf{U}_t(t-s)\mathbf{Q}^T, \mathbf{Q}\mathbf{V}(t)\mathbf{Q}^T) = \mathbf{Q}\hat{\mathbf{G}}(\mathbf{U}_t(t-s), \mathbf{V}(t))\mathbf{Q}^T \tag{19.39}$$

for all $\mathbf{Q} \in \mathscr{O}(\mathscr{V})$.

As before there is no known general solution for the condition of isotropy (19.39). However, for elastic isotropic solids, which are devoid of any memory effect, $\hat{\mathbf{G}}$ becomes a function of $\mathbf{V} = \mathbf{V}(t)$ only, and the condition (19.39) reduces to

$$\hat{\mathbf{G}}(\mathbf{Q}\mathbf{V}\mathbf{Q}^T) = \mathbf{Q}\hat{\mathbf{G}}(\mathbf{V})\mathbf{Q}^T. \tag{19.40}$$

By using the theorem of Rivlin and Ericksen, we obtain

$$\hat{\mathbf{G}}(\mathbf{V}) = \varphi_0\mathbf{I} + \varphi_1\mathbf{V} + \varphi_2\mathbf{V}^2, \tag{19.41}$$

where $\varphi_0, \varphi_1, \varphi_2$ are functions of the principal invariants of $\mathbf{V}$. Consequently, the constitutive equation of an isotropic elastic solid has the representation

$$\mathbf{T} = \psi_0 \mathbf{I} + \psi_1 \mathbf{V} + \psi_2 \mathbf{V}^2, \tag{19.42}$$

provided that the reference configuration $\varkappa$ is undistorted.

In continuum mechanics there are many other representation theorems which characterize the constitutive equations of various types of materials subject to different kinds of assumptions. The theorems established in this section are just some important examples. There are many types of subgroups of the special linear group $\mathscr{SL}(\mathscr{V})$; we can formulate a representation problem for each choice of $\mathscr{G}$. Our examples here illustrate a fairly general procedure for finding a representation.

20. The Energy Principle for Hyperelastic Materials

In Section 15 we introduced the concept of kinetic energy E in the motion of a body manifold $\mathscr{B}$. We have not discussed the balance of energy, however, because first, for a deformable body the kinetic energy is only a part of the total energy possessed by the body, and second, the mechanical power of the forces acting on the body is also only a part of the total energy flux. As a result, the rate of change of the kinetic energy need not be balanced by the power of the forces as in analytical mechanics.

It is well known in physics that energy may take many different forms, the most obvious examples of nonmechanical forms of energy being heat, electromagnetic energy, and chemical energy. Since all forms of energy are interchangeable, there is a general balance principle only for the rate of change of the total energy and the total energy flux. This general principle is usually called the *first law of thermodynamics*. In this section we have no intention to venture outside the domain of continuum mechanics. For this reason we shall limit ourselves to a very special type of energy balance, which is valid only for a certain class of elastic materials under certain physical restrictions.

Specifically, we assume that there is a balance between the mechanical power induced by the forces acting on the body and the rate of change of the total mechanical energy, which is assumed to be the sum of the kinetic energy E and the (elastic) stored energy S of the body. Like E, S is a measure on $\mathscr{B}$ and is given by the integral

$$S(\mathscr{P}) = \int_{\chi_t(\mathscr{P})} \varrho\varepsilon \, dx, \qquad \mathscr{P} \subset \mathscr{B} \tag{20.1}$$

at any instant t in a motion χ_t of $\mathscr{B}$. Then the stored energy density ε, like the stress tensor $\mathbf{T}$, is given by a constitutive equation of the form

$$\varepsilon(x, t) = \varphi(\mathbf{F}(\mathbf{X}, t), X, \varkappa). \tag{20.2}$$

Notice that only the present value $\mathbf{F}(\mathbf{X}, t)$ of the deformation gradient enters into the argument of φ, since we have assumed that the material is elastic. As before we shall suppress the notations $\mathbf{x}$, t, $\mathbf{X}$, X, and $\varkappa$ from the constitutive equation (20.2). Thus we write

$$\varepsilon = \varphi(\mathbf{F}). \tag{20.3}$$

Under the preceding assumption the special form of energy balance may be stated as follows:

Energy Principle for Hyperelastic Materials. Relative to an inertial frame the rate of change of the mechanical energy is equal to the power of the forces,

$$\frac{d}{dt}\int_{\chi_t(\mathscr{P})} \varrho(\varepsilon + \tfrac{1}{2}\mathbf{v}\cdot\mathbf{v})\,dx = \int_{\chi_t(\mathscr{P})} \varrho\mathbf{b}\cdot\mathbf{v}\,dx + \int_{\partial\chi_t(\mathscr{P})} \mathbf{t}\cdot\mathbf{v}\,d\sigma, \tag{20.4}$$

where $\mathscr{P}$ is an arbitrary subbody of $\mathscr{B}$.

Note. We shall prove that an elastic material satisfies the balance principle (20.4) if and only if its response function $\mathbf{G}$ takes on a certain special form. In continuum mechanics an elastic material with such a form of response function is known as a *hyperelastic material.* In the context of continuum thermodynamics the special balance principle (20.4) is valid under two types of physical restrictions: First, the isothermal condition, which requires the temperature be held constant throughout the motion. Under this condition the free energy plays the role of the stored energy in (20.4). Second, the isentropic condition, which requires the entropy be held constant throughout the motion. Under this condition the internal energy plays the role of the stored energy. We mention these conditions here merely to point out the fact that (20.4) is not a general physical principle.

Now using the stress principle (16.4) and the balance of moment of momentum (17.34), we can rewrite the last term of (20.4) as

$$\int_{\partial\chi_t(\mathscr{P})} \mathbf{t}\cdot\mathbf{v}\,d\sigma = \int_{\partial\chi_t(\mathscr{P})} \mathbf{Tn}\cdot\mathbf{v}\,d\sigma = \int_{\partial\chi_t(\mathscr{P})} \mathbf{Tv}\cdot\mathbf{n}\,d\sigma, \tag{20.5}$$

which may be converted into a volume integral by means of the divergence theorem

$$\int_{\partial \chi_t(\mathscr{P})} \mathbf{T}\mathbf{v} \cdot \mathbf{n}\, d\sigma = \int_{\chi_t(\mathscr{P})} \mathbf{v} \cdot \operatorname{div} \mathbf{T}\, dx + \int_{\chi_t(\mathscr{P})} \operatorname{tr}(\mathbf{T} \operatorname{grad} \mathbf{v})\, dx. \tag{20.6}$$

Substituting (20.6) into (20.4) and using the linear momentum equation (17.12), we get

$$\int_{\chi_t(\mathscr{P})} [\varrho\dot{\varepsilon} - \operatorname{tr}(\mathbf{T} \operatorname{grad} \mathbf{v})]\, dx = 0. \tag{20.7}$$

As before since $\mathscr{P}$ is arbitrary, (20.7) implies the field equation

$$\varrho\dot{\varepsilon} = \operatorname{tr}(\mathbf{T} \operatorname{grad} \mathbf{v}), \tag{20.8}$$

which is the differential form of the energy principle (20.4).

From (20.8) we see that the change of the stored energy ε is caused by the action of the stress tensor on the velocity gradient. For this reason we call the term on the right-hand side of (20.8) the *stress power*. From (14.21) the stress power is a linear function of $\dot{\mathbf{F}}$, viz.,

$$\operatorname{tr}(\mathbf{T} \operatorname{grad} \mathbf{v}) = \operatorname{tr}(\mathbf{T}\dot{\mathbf{F}}\mathbf{F}^{-1}) = \operatorname{tr}\big(\mathbf{F}^{-1}\mathbf{G}(\mathbf{F})\dot{\mathbf{F}}\big), \tag{20.9}$$

and this linear function is uniquely determined by the deformation gradient $\mathbf{F}$. From (20.3) the rate of change of the stored energy is also a linear function of $\dot{\mathbf{F}}$, viz.,

$$\varrho\dot{\varepsilon} = \operatorname{tr}\left(\varrho\left(\frac{\partial\varphi}{\partial\mathbf{F}}\right)^T \dot{\mathbf{F}}\right) = \operatorname{tr}\left(\frac{\varrho_\kappa}{\det \mathbf{F}}\left(\frac{\partial\varphi}{\partial\mathbf{F}}\right)^T \dot{\mathbf{F}}\right), \tag{20.10}$$

and this linear function is also uniquely determined by the deformation gradient $\mathbf{F}$. Now since the condition (20.8) must hold in all motions, we can regard $\mathbf{F}$ and $\dot{\mathbf{F}}$ as mutually independent variables. As a result, the linear function given by (20.9) must be the same as that given by (20.10). In other words, we have

$$\mathbf{F}^{-1}\mathbf{G}(\mathbf{F}) = \varrho\left(\frac{\partial\varphi}{\partial\mathbf{F}}\right)^T \tag{20.11}$$

or, equivalently,

$$\mathbf{G}(\mathbf{F}) = \varrho\mathbf{F}\left(\frac{\partial\varphi}{\partial\mathbf{F}}\right)^T, \tag{20.12}$$

which shows that the stored energy function $\varphi(\mathbf{F})$ is the potential for the response function $\mathbf{G}(\mathbf{F})$.

Note. The relation between the stored energy and the stress may be expressed even more clearly by using the Piola–Kirchhoff stress tensor $\mathbf{T}_\varkappa$. Indeed, from (17.7), (17.21), and (20.12) we have

$$\mathbf{T}_\varkappa = \varrho_\varkappa \frac{\partial \varphi}{\partial \mathbf{F}} = \frac{\partial}{\partial \mathbf{F}} (\varrho_\varkappa \varphi). \tag{20.13}$$

We can obtain this relation directly from the energy principle by transferring the integrals to the reference configuration $\boldsymbol{\varkappa}(B)$ also,

$$\frac{d}{dt} \int_{\varkappa(\mathscr{P})} \varrho_\varkappa(\varepsilon + \tfrac{1}{2}\mathbf{v} \cdot \mathbf{v})\, dX = \int_{\varkappa(\mathscr{P})} \varrho_\varkappa \mathbf{b} \cdot \mathbf{v}\, dX + \int_{\partial\varkappa(\mathscr{P})} \mathbf{T}_\varkappa \mathbf{N} \cdot \mathbf{v}\, d\Sigma. \tag{20.14}$$

Then by the argument as before we get the field equation

$$\varrho_\varkappa \dot{\varepsilon} = \operatorname{tr}(\mathbf{T}_\varkappa{}^T \dot{\mathbf{F}}), \tag{20.15}$$

which implies the relation (20.13).

The particular form of response function $\mathbf{G}(\mathbf{F})$ given by (20.12) is the one mentioned before. We call an elastic material having a constitutive equation of the form

$$\mathbf{T} = \varrho \mathbf{F} \left(\frac{\partial \varphi}{\partial \mathbf{F}} \right)^T \tag{20.16}$$

a *hyperelastic material.*

The constitutive equation (20.16) must obey the principle of material frame-indifference (18.3), which implies that

$$\mathbf{Q}^T \left(\frac{\partial \varphi}{\partial \mathbf{F}} \right) \bigg|_{\mathbf{QF}} = \left(\frac{\partial \varphi}{\partial \mathbf{F}} \right) \bigg|_{\mathbf{F}} \tag{20.17}$$

for all rotations $\mathbf{Q}$. Since (20.3) is also a constitutive equation, we assume that it, too, obeys the principle of material frame-indifference. Specifically, we require

$$\varphi(\mathbf{QF}) = \varphi(\mathbf{F}) \tag{20.18}$$

for all $\mathbf{Q} \in \mathscr{SO}(\mathscr{V})$. This condition reflects our understanding that ε, like the stress tensor $\mathbf{T}$, is caused by the deformations of the body. Then (20.18) asserts that the stored energy of the body is invariant when the body merely suffers only a rigid rotation. It turns out that the conditions (20.17) and (20.18) are mathematically equivalent. This result was noted originally by Noll.

Sufficiency of (20.18) for (20.17) is obvious. We simply differentiate (20.18) with respect to $\mathbf{F}$ to get (20.17). Conversely, integrating (20.17), we obtain

$$\varphi(\mathbf{QF}) = \varphi(\mathbf{F}) + \varphi(\mathbf{Q}) + \varphi(\mathbf{I}). \tag{20.19}$$

Thus it remains to show that

$$\varphi(\mathbf{Q}) = \varphi(\mathbf{I}) \tag{20.20}$$

for all $\mathbf{Q} \in \mathscr{SO}(\mathscr{V})$. We set $\mathbf{F} = \bar{\mathbf{Q}}$ in (20.19) and integrate the resulting equation over $\mathscr{SO}(\mathscr{V})$ with respect to a left-invariant volume tensor $\mathbf{E}(\bar{\mathbf{Q}})$ as explained in Section 72, IVT-2, obtaining

$$\int_{\mathscr{SO}(\mathscr{V})} \varphi(\mathbf{Q}\bar{\mathbf{Q}})\mathbf{E}(\bar{\mathbf{Q}}) = \int_{\mathscr{SO}(\mathscr{V})} \varphi(\bar{\mathbf{Q}})\mathbf{E}(\bar{\mathbf{Q}}) + (\varphi(\mathbf{Q}) - \varphi(\mathbf{I})) \int_{\mathscr{SO}(\mathscr{V})} \mathbf{E}(\bar{\mathbf{Q}}). \tag{20.21}$$

From the left invariance of $\mathbf{E}(\bar{\mathbf{Q}})$ the left-hand side of (20.21) is equal to the first term on the right-hand side. Hence (20.21) reduces to

$$(\varphi(\mathbf{Q}) - \varphi(\mathbf{I})) \int_{\mathscr{SO}(\mathscr{V})} \mathbf{E}(\bar{\mathbf{Q}}) = 0. \tag{20.22}$$

Since $\mathscr{SO}(\mathscr{V})$ is bounded, the integral of $\mathbf{E}(\bar{\mathbf{Q}})$ over the whole group is finite and nonzero [cf. (72.19) in Section 72, IVT-2]. Thus (20.22) implies (20.20).

Another general condition which must be satisfied by the constitutive equation (20.16) is the equation of moment of momentum (17.34). We now show that this condition is also equivalent to the condition (20.18).

Indeed, by an argument similar to that in the proof of Noll's representation theorem in Section 18, we see that a general solution of (20.18) is

$$\varphi(\mathbf{F}) = \varphi(\mathbf{U}) = \varphi((\mathbf{F}^T\mathbf{F}))^{1/2}. \tag{20.23}$$

Differentiating this condition with respect to $\mathbf{F}$, we get

$$\frac{\partial\varphi}{\partial\mathbf{F}} = \mathbf{F}\frac{\partial\varphi}{\partial\mathbf{U}}, \tag{20.24}$$

where we have used the symmetry condition

$$\frac{\partial\varphi}{\partial\mathbf{U}} = \left(\frac{\partial\varphi}{\partial\mathbf{U}}\right)^T, \tag{20.25}$$

which follows from the symmetry of **U**. Substituting (20.24) into (20.12), we obtain

$$\mathbf{G}(\mathbf{F}) = \varrho \mathbf{F} \frac{\partial \varphi}{\partial \mathbf{U}} \mathbf{F}^T \tag{20.26}$$

which is symmetric by virtue of the condition (20.25).

Conversely, since the chain rule

$$\varrho \frac{d}{d\tau} \varphi(\mathbf{F}(\tau)) = \operatorname{tr}\left(\varrho \mathbf{F}\left(\frac{\partial \varphi}{\partial \mathbf{F}}\right)^T \mathbf{F}'\mathbf{F}^{-1}\right) = \operatorname{tr}(\mathbf{G}(\mathbf{F})\mathbf{F}'\mathbf{F}^{-1}) \tag{20.27}$$

holds for any curve $\mathbf{F} = \mathbf{F}(\tau)$ in $\mathscr{GL}(\mathscr{V})$, by choosing the particular curve

$$\mathbf{F}(\tau) = \mathbf{Q}(\tau)\mathbf{F}_0 \tag{20.28}$$

with a skew-symmetric velocity gradient

$$\mathbf{F}'\mathbf{F}^{-1} = \mathbf{Q}'\mathbf{Q}^T = \mathbf{\Omega} = -\mathbf{\Omega}^T, \tag{20.29}$$

we get

$$\varrho \frac{d}{d\tau} \varphi(\mathbf{Q}(\tau)\mathbf{F}_0) = \operatorname{tr}(\mathbf{G}(\mathbf{F})\mathbf{\Omega}) = 0. \tag{20.30}$$

Thus the value of φ remains constant on each curve of the form (20.28). This condition implies (20.18) immediately, since the rotation group $\mathscr{SO}(\mathscr{V})$ is connected.

The foregoing results may be summarized as follows:

Theorem (Noll[6]). The following three conditions are mathematically equivalent:

(i) The value **T** of the response function **G(F)** given by (20.12) is always a symmetric tensor.

(ii) The response function obeys the principle of material frame-indifference (20.17).

(iii) The stored energy function obeys the principle of material frame-indifference (20.18).

[6] W. Noll, On the continuity of the solid and fluid states, *Journal of Rational Mechanics and Analysis*, Vol. 4, pp. 3–81, 1955. Reprinted in *Continuum Mechanics*, International Science Review Series, Vol. 8, Edited by C. Truesdell, Gordon and Breach, New York, 1965.

Before closing this section, we remark that hyperelastic materials, which are defined by constitutive equations of the form (20.16), form only a proper subset of elastic materials in general. We can characterize that subset by a condition of integrability for the existence of the potential φ. Indeed, from (20.27) we must have

$$\oint \frac{1}{\varrho} \operatorname{tr}\big(\mathbf{G}(\mathbf{F})\mathbf{F}'\mathbf{F}^{-1}\big)\, d\tau = 0 \tag{20.31}$$

for any closed circuit $\mathbf{F} = \mathbf{F}(\tau)$. If this condition holds, we can determine the potential φ by

$$\varphi(\mathbf{F}) = \int_{\mathbf{I}}^{\mathbf{F}} \frac{1}{\varrho} \operatorname{tr}\big(\mathbf{G}(\mathbf{F})\mathbf{F}'\mathbf{F}^{-1}\big)\, d\tau + \varphi(\mathbf{I}), \tag{20.32}$$

where the integral is taken along any curve $\mathbf{F}(\tau)$ joining $\mathbf{I}$ to $\mathbf{F}$, and where $\varphi(\mathbf{I})$ is an arbitrary constant.

21. Internal Constraints

So far we have formulated the basic governing principles for the motions of a body manifold $\mathscr{B}$ which is free of any constraint. Given any motion of $\mathscr{B}$, we can determine the stress tensor field $\mathbf{T} = \mathbf{T}(\mathbf{x}, t)$ by using the constitutive equation. Then from the stress tensor field we can determine the body force field by using the linear momentum equation. We generally regard the body force as an external quantity, which is not subject to any *a priori* restrictions. Consequently, our theory allows the motions of $\mathscr{B}$ to be arbitrary.

In this section we extend our formulation to bodies which are subject to certain constraints. Previously we have explained that a configuration of a body manifold $\mathscr{B}$, unlike that of a system $\mathscr{Q}$ of particles and rigid bodies, cannot be characterized by a finite set of parameters. Hence the free configuration space, which may be taken as the collection of all configurations of $\mathscr{B}$, is not a finite-dimensional manifold. We can define a holonomic constraint on $\mathscr{B}$ by requiring that the configurations of $\mathscr{B}$ be restricted to stay on a certain constraint surface in the free configuration space. Then in general the constraint surface is also infinite dimensional.

Note. We can regard a rigid body as a body manifold which is subject to the constraint of rigidity. For this special case the constraint surface is finite dimensional, of course. As we shall see, the governing principles for this special case reduce precisely to those introduced in Section 4.

In continuum mechanics we restrict our attention to the following special type of infinite-dimensional constraint suffaces which may be described by means of certain finite-dimensional entities. First, we choose a fixed reference configuration $\boldsymbol{\varkappa}(\mathscr{B})$ to represent the body manifold $\mathscr{B}$. As has always been the case, this reference configuration may or may not be an actual configuration of $\mathscr{B}$ at a particular instant in a motion. Consequently, $\boldsymbol{\varkappa}$ need not itself satisfy the condition of constraint, defined as follows:

We specify a smooth surface

$$\mathscr{M} = \mathscr{M}(X, \boldsymbol{\varkappa}) \subset \mathscr{GL}(\mathscr{V})^{\dagger} \tag{21.1}$$

for each body point $X \in \mathscr{B}$. Then we require the configurations of $\mathscr{B}$ to be restricted in such a way that the deformation gradient $\mathbf{F}$ relative to $\boldsymbol{\varkappa}$ must satisfy the condition

$$\mathbf{F}(\mathbf{X}, t) \in \mathscr{M}(X, \boldsymbol{\varkappa}) \tag{21.2}$$

for all $\mathbf{X} \in \boldsymbol{\varkappa}(\mathscr{B})$ and for all $t \in \mathscr{R}$. In particular, the deformation history $\mathbf{F}(\mathbf{X}, t - s)$, $s \in [0, \infty)$, must stay in the surface $\mathscr{M}(X, \boldsymbol{\varkappa})$, so the domain of the response functional is restricted.

Note. The surface $\mathscr{M}$ may or may not contain the identity element $\mathbf{I}$ of $\mathscr{GL}(\mathscr{V})^{\dagger}$. Since the deformation gradient from $\boldsymbol{\varkappa}$ to itself is $\mathbf{I}$, $\boldsymbol{\varkappa}$ satisfies the condition of constraint (21.2) if and only if $\mathscr{M}$ happens to contain $\mathbf{I}$ for all $X \in \mathscr{B}$.

Now in accord with our understanding of the constitutive equation, the stress tensor is due to the internal reaction of the body toward the deformation history. Since the deformation is subject to the constraint condition (21.2), we expect that there be an additional internal reaction which affects the motion of $\mathscr{B}$ in such a way that (21.2) is maintained at all times. Hence the stress tensor $\mathbf{T} = \mathbf{T}(\mathbf{x}, t)$ consists of two parts:

$$\mathbf{T} = \mathbf{N} + \mathbf{G}\big(\mathbf{F}(t - s), s \in [0, \infty)\big), \tag{21.3}$$

where $\mathbf{N} = \mathbf{N}(\mathbf{x}, t)$ denotes the stress due to the internal reaction of the body toward the constraint, while $\mathbf{G}\big(\mathbf{F}(t - s), s \in [0, \infty)\big)$ is the stress due to the response of the body toward the deformation history, which is subject to the condition (21.2).

As remarked in analytical mechanics, the reaction **N** toward the constraint is generally an unknown quantity. Hence we must make some assumptions about its mathematical properties. In this regard we are guided by the idea as before: We require that **N** produce no power in any motion consistent with the constraint. This idea may be formulated in the following way:

In Section 20 we introduced the concept of stress power within the context of a hyperelastic material body. Now for a simple material body in general the stress power plays the role of the density of the excess mechanical power in any motion of $\mathscr{B}$. Indeed, from the linear momentum equation and the moment of momentum equation we have the identity

$$\int_{\chi_t(\mathscr{P})} \operatorname{tr}(\mathbf{T} \operatorname{grad} \mathbf{v})\, dx$$
$$= \int_{\chi_t(\mathscr{P})} \varrho \mathbf{b} \cdot \mathbf{v}\, dx + \int_{\partial\chi_t(\mathscr{P})} \mathbf{t} \cdot \mathbf{v}\, d\sigma - \frac{d}{dt} \int_{\chi_t(\mathscr{P})} \varrho \mathbf{v} \cdot \mathbf{v}\, dx \tag{21.4}$$

for any subbody $\mathscr{P}$ of $\mathscr{B}$. By virtue of this identity we see that **N** produces no excess mechanical power if and only if

$$\operatorname{tr}(\mathbf{N} \operatorname{grad} \mathbf{v}) = \operatorname{tr}(\mathbf{N}\dot{\mathbf{F}}\mathbf{F}^{-1}) = 0 \tag{21.5}$$

for all deformation histories $\mathbf{F}(t - s)$ consistent with the constraint (21.2).

Note. Strictly speaking, the excess mechanical power of **N** is given by the integral of $\operatorname{tr}(\mathbf{N}^T \operatorname{grad} \mathbf{v})$, since we do not know whether **N** is symmetric or not. The condition

$$\operatorname{tr}(\mathbf{N}^T \operatorname{grad} \mathbf{v}) = \operatorname{tr}(\mathbf{N}^T\dot{\mathbf{F}}\mathbf{F}^{-1}) = 0, \tag{21.6}$$

however, implies that

$$\mathbf{N} = \mathbf{N}^T \tag{21.7}$$

under a very natural assumption. We generally require that the surface $\mathscr{M}$ be closed with respect to superposed rigid rotation; i.e., if $\mathbf{F} \in \mathscr{M}$, then $\mathbf{QF} \in \mathscr{M}$ for all rotations **Q**. This assumption is reasonable, since the condition of constraint (21.2) is meant to be a restriction on the internal deformation, not a restriction on the position of $\chi_t(\mathscr{B})$ in space. Hence if the configuration χ_t is consistent with the constraint, then the configuration $\mathbf{Q} \circ \boldsymbol{\chi}_t$, obtained from $\boldsymbol{\chi}_t$ by any superposed rigid rotation $\mathbf{Q}$, is also. We call this assumption the *condition of frame-indifference* for the constraint. Under this condition, if we choose any point $\mathbf{F}_0 \in \mathscr{M}$ and consider a curve passing through it,

$$\mathbf{F}(\tau) = \mathbf{Q}(\tau)\mathbf{F}_0, \tag{21.8}$$

where $\mathbf{Q}(0) = \mathbf{I}$ and $\dot{\mathbf{Q}}(0) = \boldsymbol{\Omega}$, then the condition (21.6) implies that

$$\operatorname{tr}(\mathbf{N}^T\boldsymbol{\Omega}) = 0. \tag{21.9}$$

From (21.9) we see that $\mathbf{N}$ is orthogonal to all skew-symmetric tensors $\boldsymbol{\Omega}$ relative to the standard inner product on the space of second-order tensors [cf. equation (35.50) in Section 35, IVT-1]. As a result, $\mathbf{N}$ is necessarily a symmetric tensor, and thus the conditions (21.5) and (21.6) are equivalent.

Conversely, the symmetry of $\mathbf{N}$ and (21.5) or (21.6) imply also the condition of frame-indifference. Indeed, we can consider an arbitrary curve of the form (21.8) and obtain as before the condition

$$\operatorname{tr}(\mathbf{N}\dot{\mathbf{F}}\mathbf{F}^{-1}) = \operatorname{tr}(\mathbf{N}\dot{\mathbf{Q}}\mathbf{Q}^T) = 0, \tag{21.10}$$

since the tensor $\dot{\mathbf{Q}}\mathbf{Q}^T$ is skew symmetric. As a result, we can extend the surface $\mathscr{M}$ into a surface satisfying the condition of frame-indifference without violating the condition (21.5).

The condition (21.5) may be solved explicitly by using Lagrange's multipliers. That is, $\mathbf{N}$ is expressible as a linear combination of a basis for the orthogonal complement of the tangent space of $\mathscr{M}$ at the present deformation gradient $\mathbf{F}(t)$. For example, if $\mathscr{M}$ is a hypersurface which is characterized by a single algebraic equation of the form

$$\mu(\mathbf{F}) = 0, \tag{21.11}$$

then by taking the derivative of μ on any curve $\mathbf{F} = \mathbf{F}(t)$ which satisfies the constraint (21.11), we get

$$\operatorname{tr}\left[\left(\frac{\partial \mu}{\partial \mathbf{F}}\right)^T \dot{\mathbf{F}}\right] = \operatorname{tr}\left[\mathbf{F}\left(\frac{\partial \mu}{\partial \mathbf{F}}\right)^T \dot{\mathbf{F}}\mathbf{F}^{-1}\right] = 0. \tag{21.12}$$

Comparing this equation with (21.5), we obtain

$$\mathbf{N} = \lambda \mathbf{F}\left(\frac{\partial \mu}{\partial \mathbf{F}}\right)^T, \tag{21.13}$$

where λ is a Lagrange's multiplier.

Note. From (21.13) and Noll's theorem in Section 20 we see also that the condition (21.7) is equivalent to the condition of frame-indifference, which in this case means that

$$\mu(\mathbf{QF}) = \mu(\mathbf{F}) \tag{21.14}$$

for all $\mathbf{Q} \in \mathscr{SO}(\mathscr{V})$. It follows that μ reduces to a function of the right stretch tensor $\mathbf{U}$ of $\mathbf{F}$ only, viz.,

$$\mu(\mathbf{F}) = \mu(\mathbf{U}). \tag{21.15}$$

Using the representation (21.15), we can rewrite the tensor $\mathbf{N}$ as

$$\mathbf{N} = \lambda \mathbf{F} \frac{\partial \mu}{\partial \mathbf{U}} \mathbf{F}^T, \tag{21.16}$$

which is clearly symmetric.

In general if $\mathscr{M}$ is a surface of lower dimension, say, $9 - k$, then it may be characterized by a system of algebraic equations, viz., $\mathbf{F} \in \mathscr{M}$ if and only if

$$\mu_a(\mathbf{F}) = 0, \qquad a = 1, \ldots, k. \tag{21.17}$$

In this case the representation (21.13) is replaced by

$$\mathbf{N} = \lambda_1 \mathbf{F}\left(\frac{\partial \mu_1}{\partial \mathbf{F}}\right)^T + \cdots + \lambda_k \mathbf{F}\left(\frac{\partial \mu_k}{\partial \mathbf{F}}\right)^T = \lambda_a \mathbf{F}\left(\frac{\partial \mu_a}{\partial \mathbf{F}}\right)^T. \tag{21.18}$$

Also, if the functions μ_a, $a = 1, \ldots, k$, obey the condition of frame-indifference, then we have the symmetric representation

$$\mathbf{N} = \lambda_a \mathbf{F} \frac{\partial \mu_a}{\partial \mathbf{U}} \mathbf{F}^T, \tag{21.19}$$

where a is summed from 1 to k.

We now consider some important special cases of internal constraints:

(i) *Incompressibility*. For this constraint we require that the determinant of $\mathbf{F}$ remain fixed in all motions. We can express this constraint by an equation of the form (21.11), where $\mu(\mathbf{F}) = \det \mathbf{F} - c$, $c > 0$. Using this function, we obtain directly from (21.13) that the orthogonal complement of $\mathscr{M}$ is the 1-dimensional subspace spanned by the identity tensor $\mathbf{I}$. Thus $\mathbf{N}$ is a hydrostatic stress, viz.,

$$\mathbf{N} = -p\mathbf{I}, \tag{21.20}$$

where p denotes the pressure. Substituting (21.20) into (21.3), we see that the constitutive equation for an incompressible simple material is

$$\mathbf{T} = -p\mathbf{I} + \mathbf{G}(\mathbf{F}(t - s), s \in [0, \infty)). \tag{21.21}$$

(ii) *Inextensibility*. For this constraint we require that the deformation gradient $\mathbf{F}$ preserve the length of a particular vector $\mathbf{e}$ in all motions. We can express the constraint again by an equation of the form (21.11), where $\mu(\mathbf{F}) = \| \mathbf{Fe} \| - c$, $c > 0$. Using this function, we obtain from (21.13) the result that the orthogonal complement of $\mathscr{M}$ at any $\mathbf{F} \in \mathscr{M}$ is spanned by the tensor $\mathbf{Fe} \otimes \mathbf{Fe}$. Thus $\mathbf{N}$ is a uniaxial tension in the direction of $\mathbf{Fe}$, viz.,

$$\mathbf{N} = \lambda \mathbf{Fe} \otimes \mathbf{Fe}. \tag{21.22}$$

From this representation we see that the constitutive equation of an inextensible simple material is of the form

$$\mathbf{T} = \lambda \mathbf{Fe} \otimes \mathbf{Fe} + \mathbf{G}\big(\mathbf{F}(t - s), s \in [0, \infty)\big). \tag{21.23}$$

(iii) *Rigidity*. This is a degenerate constraint for which $\mathscr{M}$ is a 3-dimensional surface formed by tensors of the form $\mathbf{QF}_0$, where $\mathbf{F}_0$ is fixed and $\mathbf{Q} \in \mathscr{SO}(\mathscr{V})$. For this case the orthogonal complement at any point of $\mathscr{M}$ is 6-dimensional, so that it consists of all symmetric tensors. Thus for a rigid body the stress tensor is entirely independent of the motion.

The preceding three examples illustrate clearly the general properties of the constraint stress $\mathbf{N}$, namely, $\mathbf{N}$ is an appropriate tensor whose presence in $\boldsymbol{\chi}_t$ prevents the body $\mathscr{B}$ from violating the given constraint. The precise value of $\mathbf{N}$ depends on the external body forces and the boundary tractions. Specifically, given any motion of $\mathscr{B}$ consistent with the constraint, we determine first the value $\mathbf{G}\big(\mathbf{F}(t - s), s \in [0, \infty)\big)$ due to the response. Then from the linear momentum equation we determine the combination $\varrho\mathbf{b} + \operatorname{div}\mathbf{N}$. In particular, two external body force fields $\mathbf{b}$ and $\bar{\mathbf{b}}$ differing only by

$$\mathbf{b} - \bar{\mathbf{b}} = \frac{1}{\varrho} \operatorname{div}\mathbf{K}, \tag{21.24}$$

where $\mathbf{K}$ is orthogonal to the constraint surface $\mathscr{M}$, are consistent with the same motion.

In particular, for a rigid body any two force systems which have the same resultant force and the same resultant moment are equivalent, since in the theory of partial differential equations it is known that the boundary value problem

$$\begin{aligned} \operatorname{div} \mathbf{K} &= \mathbf{f} \quad \text{in} \quad \mathscr{D}, \\ \mathbf{K}\mathbf{n} &= \mathbf{g} \quad \text{on} \quad \partial\mathscr{D}, \end{aligned} \tag{21.25}$$

has a solution if and only if **f** and **g** satisfy the compatibility conditions:

$$\int_{\mathscr{D}} \mathbf{f}\, dx + \int_{\partial\mathscr{D}} \mathbf{g}\, d\sigma = \mathbf{0} \tag{21.26}$$

and

$$\int_{\mathscr{D}} \mathbf{x}\times\mathbf{f}\, dx + \int_{\partial\mathscr{D}} \mathbf{x}\times\mathbf{g}\, d\sigma = \mathbf{0}. \tag{21.27}$$

Indeed, we can regard (21.25) as a traction boundary value problem in the classical theory of linear elasticity by identifying **K** as the stress tensor of any linearly elastic material (say, with a positive definite stored energy function so as to assure the existence[7] of a solution).

[7] Compare G. Fichera, Existence theorems in elasticity, in *Handbuch der Physik*, Edited by C. Truesdell, Vol. VIa/2, Springer-Verlag, Berlin, 1972.

4

Some Topics in the Statics and Dynamics of Material Bodies

There is a vast literature in continuum mechanics on topics ranging from the classical theories of hydrodynamics and linear elasticity to the modern theories of materials with memory effects and dislocations. In this chapter we shall consider only three topics for the purpose of illustrating the basic principles developed in the preceding chapter. The topics are: the theory of viscometric flows of simple fluids, the universal solutions of isotropic elastic solids, and the mathematical formulation of continuous distributions of dislocations in elastic bodies.

22. Homogeneous Simple Material Bodies

Noll's concept of a simple material, which we have summarized in Section 18, is basically a mathematical model for the mechanical response of a single body point X in a body manifold $\mathscr{B}$. We may, of course, apply this model to each and every point of $\mathscr{B}$. Then we obtain in a very natural way a mathematical model for the mechanical response of the whole body $\mathscr{B}$. Thus we may regard a simple material body as a body manifold such that each of its body points is equipped with a constitutive equation of the form (18.1) relative to the particular reference configuration $\boldsymbol{\varkappa}$ of $\mathscr{B}$. That is,

$$\mathbf{T}(\mathbf{x}, t) = \mathbf{G}\big(\mathbf{F}(\mathbf{X}, t-s), s \in [0, \infty), X, \boldsymbol{\varkappa}\big) \tag{22.1}$$

for all $X \in \mathscr{B}$.

In Section 18 we have pointed out that the response functional $\mathbf{G}$ on the right-hand side of (22.1) generally depends on the body point $X \in \mathscr{B}$ and on the reference configuration $\boldsymbol{\varkappa}$. This remark means that the mechanical response of different body points, taken with respect to deformation histories relative to the reference configuration $\boldsymbol{\varkappa}$, need not be the same for the whole body $\mathscr{B}$. In general the response functionals of different body points of $\mathscr{B}$ are entirely independent of one another. As a result, in order to specify a simple material body $\mathscr{B}$ we must assign a field of response functionals on the domain $\boldsymbol{\varkappa}(\mathscr{B})$. Mathematically, such a field has a rather complicated structure. In this section we develop first a theory for the simplest kind of bodies, namely, homogeneous bodies.

A *homogeneous simple material body* is a body manifold $\mathscr{B}$ such that there exists a reference configuration $\boldsymbol{\varkappa}$ relative to which the response functional $\mathbf{G}$ is independent of the body point. We call this particular reference configuration a *homogeneous configuration* of $\mathscr{B}$. In this configuration the body points may be regarded as being in the same state, since their response functionals are identical to one another.

It should be noted that a homogeneous body may have homogeneous configurations as well as inhomogeneous configurations. Indeed, suppose that $\boldsymbol{\varkappa}$ is a homogeneous configuration of $\mathscr{B}$. Then from the transformation law (18.24) another configuration $\hat{\boldsymbol{\varkappa}}$ of $\mathscr{B}$ remains a homogeneous one if and only if

$$\mathbf{G}\big(\mathbf{F}(t-s)\mathbf{K}(\mathbf{X}), \boldsymbol{\varkappa}\big) = \mathbf{G}\big(\mathbf{F}(t-s)\mathbf{K}(\mathbf{Y}), \boldsymbol{\varkappa}\big) \tag{22.2}$$

for all deformation histories $\mathbf{F}(t-s)$ and for all $\mathbf{X}, \mathbf{Y} \in \boldsymbol{\varkappa}(\mathscr{B})$, where $\mathbf{K}$ denotes the deformation gradient field from $\boldsymbol{\varkappa}$ to $\hat{\boldsymbol{\varkappa}}$ as before. Comparing (22.2) with (18.27), we see that a necessary and sufficient condition for (22.2) is

$$\mathbf{K}(\mathbf{X})^{-1}\mathbf{K}(\mathbf{Y}) \in \mathscr{G}(\boldsymbol{\varkappa}) \tag{22.3}$$

for all $\mathbf{X}, \mathbf{Y} \in \boldsymbol{\varkappa}(\mathscr{B})$, where $\mathscr{G}(\boldsymbol{\varkappa})$ denotes the symmetry group relative to $\boldsymbol{\varkappa}$. If $\mathbf{K}$ fails to satisfy the condition (22.3), then $\hat{\boldsymbol{\varkappa}}$ is an inhomogeneous configuration of $\mathscr{B}$. In such a configuration the body points are not in the same state; i.e., their response functionals are not identical.

Note. The condition (22.3) is certainly satisfied when $\mathbf{K}$ is a constant field on $\boldsymbol{\varkappa}(\mathscr{B})$; i.e., when the deformation $\boldsymbol{\varphi} = \hat{\boldsymbol{\varkappa}} \circ \boldsymbol{\varkappa}^{-1}$ from $\boldsymbol{\varkappa}$ to $\hat{\boldsymbol{\varkappa}}$ is a homogeneous one. However, if $\mathscr{G}(\boldsymbol{\varkappa})$ is a large group, then there might be some inhomogeneous deformations which also satisfy the condition (22.3). For example, if $\mathscr{G}(\boldsymbol{\varkappa}) = \mathscr{SL}(\mathscr{V})$ for a simple fluid, then (22.3) holds whenever the determinant of **K** is a constant field. Clearly, there are many inhomogeneous deformations whose gradients have a constant determinant.

If a homogeneous configuration $\boldsymbol{\varkappa}$ of $\mathscr{B}$ is given, then we can determine whether or not another configuration $\hat{\boldsymbol{\varkappa}}$ is also homogeneous by the criterion (22.3). Conversely, however, if a general inhomogeneous configuration $\hat{\boldsymbol{\varkappa}}$ is given, say the constitutive equation relative to $\hat{\boldsymbol{\varkappa}}$ is

$$\mathbf{T}(\mathbf{x}, t) = \mathbf{G}(\mathbf{F}(\hat{\mathbf{X}}, t - s), s \in [0, \infty), X, \hat{\boldsymbol{\varkappa}}), \tag{22.4}$$

where the response functional $\mathbf{G}(\cdot, \hat{\boldsymbol{\varkappa}})$ relative to $\hat{\boldsymbol{\varkappa}}$ is specified, then it is a very difficult problem to determine a deformation $\hat{\boldsymbol{\varphi}} = \boldsymbol{\varkappa} \circ \hat{\boldsymbol{\varkappa}}^{-1}$ such that the configuration $\boldsymbol{\varkappa}$ is homogeneous. Indeed, from the transformation law (18.24) the deformation gradient field $\hat{\mathbf{K}}$ of $\hat{\boldsymbol{\varphi}}$ must satisfy the condition

$$\begin{aligned}\mathbf{G}(\mathbf{F}(t - s)\hat{\mathbf{K}}(\hat{\mathbf{X}}), s \in [0, \infty), X, \hat{\boldsymbol{\varkappa}})\\ = \mathbf{G}(\mathbf{F}(t - s)\hat{\mathbf{K}}(\hat{\mathbf{Y}}), s \in [0, \infty), Y, \hat{\boldsymbol{\varkappa}})\end{aligned} \tag{22.5}$$

for all $X, Y \in \mathscr{B}$ and for all deformation histories $\mathbf{F}(t - s)$, $s \in [0, \infty)$. Since the response functionals $\mathbf{G}(\cdot, X, \hat{\boldsymbol{\varkappa}})$ and $\mathbf{G}(\cdot, Y, \hat{\boldsymbol{\varkappa}})$ are generally not the same, the condition (22.5) cannot be reduced to a simple criterion like (22.3).

In fact the existence of a solution $\hat{\mathbf{K}}$ for the condition (22.5) characterizes implicitly the fact that $\mathscr{B}$ is a homogeneous body. Since the solution $\hat{\mathbf{K}}$ must be a field of deformation gradients, it has to satisfy the compatibility condition (14.9) also. We shall consider the problem of existence of $\hat{\mathbf{K}}$ in the context of a more general model later. If (22.5) fails to have a solution, then $\mathscr{B}$ is an inhomogeneous body. For such a body all configurations are inhomogeneous.

Now suppose that $\mathscr{B}$ is a homogeneous body, and let $\boldsymbol{\varkappa}$ be a homogeneous reference configuration for $\mathscr{B}$. Then the response functional **G** relative to $\boldsymbol{\varkappa}$ is independent of the body point. In principle we can test **G** in the following way: We consider arbitrary homogeneous motions of $\mathscr{B}$ of the form

$$\mathbf{x} = \mathbf{F}\mathbf{X} + \mathbf{c}, \tag{22.6}$$

where **F** and **c** depend only on t. For this motion the stress tensor field is a constant field at any instant t, since the deformation history $\mathbf{F}(t - s)$, $s \in [0, \infty)$, is independent of the position. As a result, div **T** vanishes identically, and the equation of linear momentum reduces to

$$\mathbf{a} = \ddot{\mathbf{F}}\mathbf{X} + \ddot{\mathbf{c}} = \ddot{\mathbf{F}}\mathbf{F}^{-1}(\mathbf{x} - \mathbf{c}) + \ddot{\mathbf{c}} = \mathbf{b}. \tag{22.7}$$

If such a body force can be produced, and if suitable contact forces are applied on the boundary of the body, then the homogeneous motion (22.6) is possible, and the value of the response functional $\mathbf{G}(\mathbf{F}(t - s), \boldsymbol{\varkappa})$ is simply the constant stress tensor $\mathbf{T}(t)$ on the instantaneous configuration of $\mathscr{B}$ at time t.

It is highly questionable, of course, whether a body force field $\mathbf{b}$ of the form (22.7) can be achieved in a laboratory. Also, if the simple material of $\mathscr{B}$, in fact, has an infinite memory effect, then the homogeneous motion (22.6) must be maintained over an infinite period of time, which is clearly impossible to achieve physically. Hence in application we must either restrict the deformation histories to a class, which is representable by a certain finite-dimensional space, or conjecture that the response functional has a certain special form, which can be determined completely by testing only a few deformation histories.

From (22.7) we see that a homogeneous motion can be maintained by suitable boundary forces alone if and only if both $\ddot{\mathbf{F}}$ and $\ddot{\mathbf{c}}$ vanish identically. Thus such a motion has the explicit representation

$$\mathbf{x} = \mathbf{F}_1(\mathbf{I} + \mathbf{F}_2 t)\mathbf{X} + \mathbf{c}_1 t + \mathbf{c}_2, \tag{22.8}$$

where $\mathbf{F}_1$, $\mathbf{F}_2$, $\mathbf{c}_1$, $\mathbf{c}_2$ are constant. For this motion the deformation history is of the form

$$\mathbf{F}(t - s) = \mathbf{F}_1(\mathbf{I} + \mathbf{F}_2(t - s)), \qquad s \in [0, \infty). \tag{22.9}$$

Since the deformation gradient must have a positive determinant, the tensors $\mathbf{F}_1$ and $\mathbf{F}_2$ in the preceding two equations are not entirely arbitrary. Indeed, in order that the motion be nonsingular for all t, it is necessary and sufficient that $\det \mathbf{F}_1$ be positive, and that all real roots of the characteristic polynomial of $\mathbf{F}_2$ be zero. (The characteristic polynomial is defined in Section 26, IVT-1.)

The accelerationless homogeneous motions (22.8) enjoy the following important property: They are motions which can be maintained by boundary forces alone for all homogeneous bodies, regardless of what simple materials the bodies are made up of. Because they have this property, we call these motions *universal solutions* for all simple materials. It is clear that no other motion enjoys the same property. Among homogeneous motions those satisfying the condition $\mathbf{b} = \mathbf{0}$ are given by (22.8). On the other hand, an inhomogeneous motion cannot possibly be a universal solution for all simple materials, since the linear momentum equation requires that

$$\operatorname{div} \mathbf{G}\big(\mathbf{F}(\mathbf{X}, t - s), s \in [0, \infty), \boldsymbol{\varkappa}\big) = \varrho \mathbf{a}, \tag{22.10}$$

which cannot always hold for an arbitrary response functional $\mathbf{G}$.

Note. Strictly speaking, the response functional $\mathbf{G}$ in (22.10) is not entirely arbitrary, since it must satisfy the condition of frame-indifference. However, from Noll's representation theorem, we know that the restriction of $\mathbf{G}$ to stretch histories is arbitrary. Since the field of stretch histories of an inhomogeneous motion must be an inhomogeneous field (proof?), the values of $\mathbf{G}$ in (22.10) are, indeed, arbitrary.

Choosing $\mathbf{F}_2 = \mathbf{0}$ and $\mathbf{c}_1 = \mathbf{0}$ in (22.8), we see that all homogeneous static deformations

$$\mathbf{x} = \mathbf{F}_1\mathbf{X} + \mathbf{c}_2, \qquad \det \mathbf{F}_1 > 0 \tag{22.11}$$

are universal solutions. Similarly, a simple shearing flow of the form

$$\mathbf{x} = (\mathbf{I} + \mathbf{K}t)\mathbf{X}, \tag{22.12}$$

where $\mathbf{K}$ has the component matrix

$$[\mathbf{K}] = \begin{bmatrix} 0 & 0 & 0 \\ k & 0 & 0 \\ 0 & 0 & 0 \end{bmatrix}, \tag{22.13}$$

is also a universal solution. Since the present deformation gradient in the motion (22.12) is

$$\mathbf{F}(t) = \mathbf{I} + \mathbf{K}t, \tag{22.14}$$

which depends on t, the stress tensor $\mathbf{T}(t)$ is generally a time-dependent homogeneous field on the present configuration. These examples illustrate some of the general features of the universal solutions (22.8).

The concept of universal solutions for all simple materials may be generalized in an obvious way to the concept of universal solutions for a specific class of simple materials, e.g., the class of all simple fluids, the class of elastic materials, the class of isotropic elastic solids, etc. Thus a universal solution for simple fluids is a motion which can be maintained by boundary forces alone for all homogeneous bodies made up of simple fluids. In general, of course, a smaller class of simple materials always gives rise to a bigger family of universal solutions. In this sense the accelerationless homogeneous motions given by (22.8) are contained in all families of universal solutions. In continuum mechanics the problem of finding universal solutions for a specific class of materials is very important. We shall discuss this problem in detail for several classes of materials in later sections of this chapter.

Having considered homogeneous bodies made up of simple materials in general, we now turn our attention to bodies with internal constraints, the most important constraint being incompressibility. We call $\mathscr{B}$ an *incompressible simple material body* if the body points of $\mathscr{B}$ are all subject to the constraint of incompressibility as defined in Section 21. We choose the reference configuration $\boldsymbol{\varkappa}$ to be consistent with the constraint, so that the deformations relative to $\boldsymbol{\varkappa}$ must be isochoric; i.e.,

$$\det \mathbf{F}(\mathbf{X}, t) = 1 \tag{22.15}$$

for all $\mathbf{X} \in \boldsymbol{\varkappa}(\mathscr{B})$ and for all times t. The constitutive equation now takes the form

$$\mathbf{T}(\mathbf{x}, t) = -p\mathbf{I} + \mathbf{G}(\mathbf{F}(\mathbf{X}, t - s), s \in [0, \infty), X, \boldsymbol{\varkappa}), \tag{22.16}$$

where the domain of the response functional $\mathbf{G}$ contains isochoric histories only.

Since the pressure p in the constitutive equation (22.15) is entirely independent of the deformation history, we can adjust the value of the response functional $\mathbf{G}$ in such a way that

$$\operatorname{tr} \mathbf{G}(\mathbf{F}(\mathbf{X}, t - s), s \in [0, \infty), X, \boldsymbol{\varkappa}) = 0. \tag{22.17}$$

Under this condition both p and $\mathbf{G}$ are uniquely determined by the stress tensor $\mathbf{T}(\mathbf{x}, t)$. We shall now impose the condition (22.17) on $\mathbf{G}$. Then as before we call $\mathscr{B}$ a homogeneous body if $\mathbf{G}$ is independent of X relative to a certain reference configuration $\boldsymbol{\varkappa}$, which is called a *homogeneous configuration* of $\mathscr{B}$.

We may test the response functional $\mathbf{G}$ relative to a homogeneous reference configuration $\boldsymbol{\varkappa}$ in the following way: We consider an isochoric homogeneous motion of the form (22.6). For this motion the value of the response functional $\mathbf{G}$ is a constant field. Hence the linear momentum equation implies that

$$-\operatorname{grad} p + \varrho\mathbf{b} = \varrho(\ddot{\mathbf{F}}\mathbf{X} + \ddot{\mathbf{c}}). \tag{22.18}$$

We assume that the mass density $\varrho_{\varkappa}$ is a constant in a homogeneous configuration $\boldsymbol{\varkappa}$. Then from (22.15) and (17.7) we see that ϱ is also a constant.

Thus (22.18) may be rewritten as

$$\mathbf{b} = \ddot{\mathbf{F}}\mathbf{X} + \operatorname{grad} \zeta = \ddot{\mathbf{F}}\mathbf{F}^{-1}\mathbf{x} + \operatorname{grad} \bar{\zeta}, \tag{22.19}$$

where ζ and $\bar{\zeta}$ are arbitrary time-dependent scalar fields on the configuration $\boldsymbol{\chi}_t(\mathscr{B})$ of the motion. If we apply this body force field and a suitable contact force on the boundary, then, in principle, we can sustain the homogeneous motion, and the value of the response functional $\mathbf{G}$ is given by

$$\mathbf{G}(\mathbf{F}(t-s), s \in [0, \infty), \boldsymbol{\varkappa}) = \mathbf{T}(t) - \tfrac{1}{3}(\operatorname{tr} \mathbf{T}(t))\mathbf{I}, \tag{22.20}$$

where $\mathbf{T}(t)$ is the constant stress tensor field in the homogeneous motion at time t.

From (22.19) we see that the homogeneous motion can be maintained by boundary forces alone if and only if the vector field $\ddot{\mathbf{F}}\mathbf{F}\mathbf{x}$ is a lamellar field (cf. Section 54, IVT-2). Since the vector field depends linearly on the position $\mathbf{x}$, a necessary and sufficient condition for it to be a lamellar field is that $\ddot{\mathbf{F}}\mathbf{F}^{-1}$ is a symmetric tensor at all times t. From (22.7) the tensor $\ddot{\mathbf{F}}\mathbf{F}^{-1}$ is the acceleration gradient of the homogeneous motion. Thus the criterion for $\mathbf{b} = \mathbf{0}$ is

$$\operatorname{grad} \mathbf{a} = (\operatorname{grad} \mathbf{a})^T \tag{22.21}$$

or, equivalently,

$$\operatorname{curl} \mathbf{a} = \mathbf{0}. \tag{22.22}$$

In general a motion satisfying the condition (22.22) is called *circulation preserving*, since (22.22) is necessary and sufficient for the following condition:

$$\frac{d}{dt}\int_{\partial\chi_t(\mathscr{U})} \mathbf{v} \cdot d\mathbf{x} = \int_{\partial\chi_t(\mathscr{U})} \mathbf{a} \cdot d\mathbf{x} = \int_{\chi_t(\mathscr{U})} \operatorname{curl} \mathbf{a} \cdot \mathbf{n}\, d\sigma = 0, \tag{22.23}$$

which means that the circulation around the boundary of any oriented surface $\mathscr{U}$ in $\mathscr{B}$ is time independent.

Summarizing the preceding analysis, we obtain the following result of Coleman and Truesdell.[1]

Theorem. A homogeneous isochoric motion is possible in every homogeneous incompressible simple material body subject to boundary forces alone if and only if it is circulation preserving.

[1] B. D. Coleman and C. Truesdell, Homogeneous motions of incompressible materials, *Zeitschrift für Angewandte Mathematik und Physik*, Vol. 45, pp. 547–551, 1965.

As explained before, an inhomogeneous isochoric motion cannot possibly be maintained by boundary forces alone for all homogeneous incompressible material bodies. Hence the preceding theorem implies that the circulation-preserving homogeneous isochoric motions are the only universal solutions for all incompressible simple materials. As remarked before, these motions are contained in all families of universal solutions for all classes of incompressible simple materials. Of course, the accelerationless isochoric homogeneous motions, such as the simple shearing flow (22.12) considered before, are *ipso facto* circulation preserving.

One example of a circulation-preserving isochoric homogeneous motion in which the acceleration does not vanish is given by the steady extension flow with deformation gradient

$$\mathbf{F}(t) = \exp(\mathbf{A}t), \tag{22.24}$$

where $\mathbf{A}$ is a constant traceless symmetric tensor, say,

$$[\mathbf{A}] = \begin{bmatrix} a & 0 & 0 \\ 0 & b & 0 \\ 0 & 0 & c \end{bmatrix}, \qquad a + b + c = 0. \tag{22.25}$$

For an explanation of the exponential map of a tensor see Section 65, IVT-2. The motion (22.24) is isochoric since

$$\det \mathbf{F}(t) = \det(\exp(\mathbf{A}t)) = \exp(\operatorname{tr}(\mathbf{A}t)) = \exp(0) = 1 \tag{22.26}$$

for all $t \in \mathscr{R}$; cf. part (g), Exercise 65.1, Section 65, IVT-2.

Now for the homogeneous motion with deformation gradient $\mathbf{F}(t)$ given by (22.24) we have

$$\mathbf{F}(t)^{-1} = \exp(-\mathbf{A}t), \qquad \ddot{\mathbf{F}}(t) = \mathbf{A}^2 \exp(\mathbf{A}t). \tag{22.27}$$

Thus the acceleration gradient is a constant symmetric tensor field, viz.,

$$\operatorname{grad} \mathbf{a} = \ddot{\mathbf{F}}\mathbf{F}^{-1} = \mathbf{A}^2. \tag{22.28}$$

As a result, the criterion (22.12) is satisfied.

Note. The simple shearing flow (22.12) may be written in the form

$$\mathbf{F}(t) = \exp(\mathbf{K}t) \tag{22.29}$$

also, where $\mathbf{K}$ is given by (22.13). A motion having the general form (22.29) may be characterized by the 1-parameter group property

$$\mathbf{F}(t_1 + t_2) = \mathbf{F}(t_1)\mathbf{F}(t_2) \tag{22.30}$$

for all t_1, t_2 in $\mathscr{R}$. In particular, the relative deformation history of such a motion is given by

$$\mathbf{F}_t(t - s) = \mathbf{F}(t - s)\mathbf{F}(t)^{-1} = \mathbf{F}(t - s)\mathbf{F}(-t) = \mathbf{F}(-s), \tag{22.31}$$

which is a function of the time lapse s only independent of the present time t. For this reason we call (22.29) an *invariant motion* relative to the configuration at time $t = 0$, where $\mathbf{F}(0) = \mathbf{I}$. A general form of an invariant motion is

$$\mathbf{F}(t) = \mathbf{H}\exp(\mathbf{K}t), \tag{22.32}$$

where the reference configuration need not coincide with any instantaneous configuration of the motion.

A motion differing from an invariant motion by a superposed rigid motion is called a *motion with constant stretch history*, viz.,

$$\mathbf{F}(t) = \mathbf{Q}(t)\mathbf{H}\exp(\mathbf{K}t), \tag{22.33}$$

where $\mathbf{Q}(t) \in \mathscr{SO}(\mathscr{V})$ for all $t \in \mathscr{R}$. In continuum mechanics this general class of motions is of considerable importance. It is known that the restriction of the response functional to this class has a simple representation.[2]

23. Viscometric Flows of Incompressible Simple Fluids

In the preceding section we have considered homogeneous motions of homogeneous simple material bodies. We pointed out that there are certain special motions, called universal solutions, which may be maintained by boundary forces alone regardless of what simple materials the bodies are made up of. Since there is a great variety of simple materials involved, such special motions are limited to a few very simple motions. In this section we focus our attention on a much smaller class of materials, namely, the class of incompressible simple fluids. Also, we drop the requirement that the motion be a universal solution. Then we can consider a bigger class of motions, including some inhomogeneous ones.

[2] C.-C. Wang, A representation theorem for the constitutive equation of a simple material in motions with constant stretch history, *Archive for Rational Mechanics and Analysis*, Vol. 20, pp. 329–340, 1965.

All results of this section were obtained by Coleman and Noll,[3] who extended the earlier analysis of Rivlin[4] on viscometric flows of non-Newtonian fluids to simple fluids. Some experimental observations of these flows were made by Weissenberg.[5]

For an incompressible simple fluid the constitutive equation is

$$\mathbf{T} = -p\mathbf{I} + \mathbf{G}(\mathbf{U}_t(t-s)), \tag{23.1}$$

where $\mathbf{U}_t(t-s)$ denotes the right stretch tensor of the relative deformation gradient $\mathbf{F}_t(t-s)$, and where $\mathbf{G}$ is an isotropic functional, viz.,

$$\mathbf{G}(\mathbf{Q}\mathbf{U}_t(t-s)\mathbf{Q}^T) = \mathbf{Q}\mathbf{G}(\mathbf{U}_t(t-s))\mathbf{Q}^T \tag{23.2}$$

for all orthogonal tensors $\mathbf{Q}$. We have purposely left out the variable $\varrho(t)$ from the argument of $\mathbf{G}$, since under the constraint of incompressibility $\varrho(t)$ is a constant independent of t. Also, the relative stretch history $\{\mathbf{U}_t(t-s), s \in [0, \infty)\}$ must satisfy the constraint

$$\det \mathbf{U}_t(t-s) = 1 \tag{23.3}$$

for all $s \in [0, \infty)$.

In application it is more convenient to use the square of the stretch history as the independent variable of the response functional; i.e., we replace (23.1) by

$$\mathbf{T} = -p\mathbf{I} + \mathbf{H}(\mathbf{C}_t(t-s)), \tag{23.4}$$

where $\mathbf{C}_t(t-s)$, called the *right Cauchy–Green tensor*, is defined by

$$\mathbf{C}_t(t-s) = \mathbf{U}_t(t-s)^2 = \mathbf{F}_t(t-s)^T\mathbf{F}_t(t-s). \tag{23.5}$$

The form (23.4) is strictly equivalent to the form (23.1). Indeed, both $\mathbf{U}_t(t-s)$ and $\mathbf{C}_t(t-s)$ are positive definite and symmetric, so they deter-

(3) B. D. Coleman and W. Noll, On certain steady flows of general fluids, *Archive for Rational Mechanics and Analysis*, Vol. 3, pp. 289–303, 1959. Reprinted in *Continuum Mechanics*, International Science Review Series, Vol. 8, Edited by C. Truesdell, Gordon and Breach, New York, 1965.

(4) R. S. Rivlin, The hydrodynamics of non-Newtonian fluids, I, *Proceedings of the Royal Society of London*, Vol. A193, pp. 260–281, 1948; The hydrodynamics of non-Newtonian fluids, II, *Proceedings of the Cambridge Philosophical Society*, Vol. 45, pp. 88–91, 1949. Reprinted in *Continuum Mechanics*, International Science Review Series, Vol. 8, Edited by C. Truesdell, Gordon and Breach, New York, 1965.

(5) K. Weissenberg, A continuum theory of rheological phenomena, *Nature* (London), Vol. 159, pp. 310–311, 1947.

mine each other uniquely. The form (23.4) is easier to calculate, however, because $\mathbf{C}_t(t-s)$ is a polynomial of $\mathbf{F}_t(t-s)$, while $\mathbf{U}_t(t-s)$ is not. In terms of the variable $\mathbf{C}_t(t-s)$ the condition (23.2) becomes

$$\mathbf{H}(\mathbf{Q}\mathbf{C}_t(t-s)\mathbf{Q}^T) = \mathbf{Q}\mathbf{H}(\mathbf{C}_t(t-s))\mathbf{Q}^T \tag{23.6}$$

for all $\mathbf{Q} \in \mathscr{SO}(\mathscr{V})$, and the condition (23.3) reduces to

$$\det \mathbf{C}_t(t-s) = 1. \tag{23.7}$$

We now use the representation (23.4) to calculate the stress tensor in a simple shearing flow (22.12), which we have considered in the preceding section. First, from (22.14) the relative deformation history in a simple shearing flow is given by

$$\mathbf{F}_t(t-s) = \mathbf{F}(t-s)\mathbf{F}(t)^{-1} = (\mathbf{I} + \mathbf{K}(t-s))(\mathbf{I} - \mathbf{K}t) = \mathbf{I} - \mathbf{K}s, \tag{23.8}$$

which depends only on s, as we have remarked before. Next, from (23.5) the right Cauchy–Green tensor is given by

$$\begin{aligned} \mathbf{C}_t(t-s) &= \mathbf{F}_t(t-s)^T\mathbf{F}(t-s) = (\mathbf{I} - \mathbf{K}^T s)(\mathbf{I} - \mathbf{K}s) \\ &= \mathbf{I} - (\mathbf{K} + \mathbf{K}^T)s + \mathbf{K}^T\mathbf{K}s^2, \end{aligned} \tag{23.9}$$

which also depends only on s. From (22.13) the component matrix of $\mathbf{C}_t(t-s)$ is

$$[\mathbf{C}_t(t-s)] = \begin{bmatrix} 1 + k^2s^2 & -ks & 0 \\ -ks & 1 & 0 \\ 0 & 0 & 1 \end{bmatrix}. \tag{23.10}$$

Substituting (23.9) into (23.4), we see that the value of the response functional is independent of t. Thus (23.4) reduces to

$$\mathbf{T}(\mathbf{x}, t) = -p(\mathbf{x}, t)\mathbf{I} + \mathbf{S}(k), \tag{23.11}$$

where

$$\mathbf{S}(k) = \mathbf{H}(\mathbf{I} - (\mathbf{K} + \mathbf{K}^T)s + \mathbf{K}^T\mathbf{K}s^2). \tag{23.12}$$

Now using the condition of isotropy (23.6), we can show that the component matrix of $\mathbf{S}(k)$ must be of the form

$$[\mathbf{S}(k)] = \begin{bmatrix} S_{11}(k) & S_{12}(k) & 0 \\ S_{12}(k) & S_{22}(k) & 0 \\ 0 & 0 & S_{33}(k) \end{bmatrix}, \tag{23.13}$$

where $S_{11}(k)$, $S_{22}(k)$, $S_{33}(k)$ are even functions of k, while $S_{12}(k)$ is an odd function; i.e.,

$$S_{11}(-k) = S_{11}(k), \qquad S_{22}(-k) = S_{22}(k), \qquad S_{33}(-k) = S_{33}(k) \tag{23.14}$$

and

$$S_{12}(-k) = -S_{12}(k) \tag{23.15}$$

for all $k \in \mathscr{R}$. This important result may be proved as follows:

We observe first that the particular tensor $\mathbf{C}_t(t - s)$ of the form (23.10) commutes with the orthogonal reflection $\mathbf{Q}$ in the direction of $\mathbf{e}_3$; i.e., $\mathbf{Q}$ has the component matrix

$$[\mathbf{Q}] = \begin{bmatrix} 1 & 0 & 0 \\ 0 & 1 & 0 \\ 0 & 0 & -1 \end{bmatrix}. \tag{23.16}$$

Hence from the condition of isotropy (23.10), $\mathbf{S}(k)$ commutes with $\mathbf{Q}$ also. This condition requires the component matrix of $\mathbf{S}(k)$ to have the form (23.13). Next, if we consider the orthogonal reflection $\bar{\mathbf{Q}}$ in the direction of $\mathbf{e}_1$, i.e.,

$$[\bar{\mathbf{Q}}] = \begin{bmatrix} -1 & 0 & 0 \\ 0 & 1 & 0 \\ 0 & 0 & 1 \end{bmatrix}, \tag{23.17}$$

then $\bar{\mathbf{Q}}(\mathbf{C}_t(t - s)\bar{\mathbf{Q}}^T$ has the component matrix

$$[\bar{\mathbf{Q}}\mathbf{C}_t(t - s)\bar{\mathbf{Q}}^T] = \begin{bmatrix} 1 + k^2s^2 & ks & 0 \\ ks & 1 & 0 \\ 0 & 0 & 1 \end{bmatrix}, \tag{23.18}$$

which is the same as $[\mathbf{C}_t(t - s)]$ except that k is replaced by $-k$. Thus

$$\mathbf{H}(\bar{\mathbf{Q}}\mathbf{C}_t(t - s)\bar{\mathbf{Q}}^T) = \mathbf{S}(-k). \tag{23.19}$$

However, from (23.6) $\mathbf{H}(\bar{\mathbf{Q}}\mathbf{C}_t(t - s)\bar{\mathbf{Q}}^T)$ is equal to $\bar{\mathbf{Q}}\mathbf{S}(k)\bar{\mathbf{Q}}^T$, and from (23.13) and (23.17)

$$[\bar{\mathbf{Q}}\mathbf{S}(k)\bar{\mathbf{Q}}^T] = \begin{bmatrix} S_{11}(k) & -S_{12}(k) & 0 \\ -S_{12}(k) & S_{22}(k) & 0 \\ 0 & 0 & S_{33}(k) \end{bmatrix}. \tag{23.20}$$

Combining (23.19) and (23.20), we obtain (23.14) and (23.15).

It should be noted that by the convention (22.17) the three normal stress components of $\mathbf{S}(k)$ must satisfy the condition

$$\operatorname{tr}(\mathbf{S}(k)) = S_{11}(k) + S_{22}(k) + S_{33}(k) = 0, \tag{23.21}$$

so they can be expressed in terms of two independent even functions of k, viz.,

$$\sigma_1(k) = S_{11}(k) - S_{33}(k), \qquad \sigma_2(k) = S_{22}(k) - S_{33}(k). \tag{23.22}$$

From (23.11) $\sigma_1(k)$ and $\sigma_2(k)$ are also the differences of the normal components of $\mathbf{T}(\mathbf{x}, t)$, viz.,

$$T_{11}(\mathbf{x}, t) - T_{33}(\mathbf{x}, t) = \sigma_1(k), \qquad T_{22}(\mathbf{x}, t) - T_{33}(\mathbf{x}, t) = \sigma_2(k). \tag{23.23}$$

The shear components of $\mathbf{S}(k)$ and $\mathbf{T}(\mathbf{x}, t)$ coincide of course. We denote the nonzero shear stress by

$$S_{12}(k) = T_{12}(\mathbf{x}, t) = \tau(k). \tag{23.24}$$

The functions $\sigma_1(k)$, $\sigma_2(k)$, and $\tau(k)$ characterize completely the response of an incompressible simple fluid in an arbitrary simple shearing flow.

We call σ_1 and σ_2 the *normal stress functions* and τ the *shear stress function*. We generally assume that τ is a monotonic increasing function which is positive for positive k [and negative for negative k by virtue of (23.15)]. That is, the higher the shearing rate the bigger the shear stress. This assumption is known to be consistent with experimental observations of fluids in simple shearing flows. We do not make any *a priori* assumptions on the normal stress functions.

In Section 22 we pointed out that the pressure field is a constant in a simple shearing flow, viz.,

$$\operatorname{grad} p = 0. \tag{23.25}$$

As a result, we can determine the boundary forces directly by

$$\mathbf{t} = \mathbf{T}\mathbf{n} = (-p\mathbf{I} + \mathbf{S}(k))\mathbf{n}; \tag{23.26}$$

e.g., if $\mathbf{n} = \mathbf{e}_1$, then $\mathbf{t}$ is

$$\mathbf{t} = T_{11}\mathbf{e}_1 + T_{12}\mathbf{e}_2 = T_{11}\mathbf{e}_1 + \tau(k)\mathbf{e}_2. \tag{23.27}$$

From this force we may test the shear stress function $\tau(k)$. The normal stress functions $\sigma_1(k)$ and $\sigma_2(k)$ require tests at several boundary points with independent normals. For example, the contact force at a boundary point where $\mathbf{n} = \mathbf{e}_3$ is

$$\bar{\mathbf{t}} - T_{33}\mathbf{e}_3. \tag{23.28}$$

Then $\sigma_1(k)$ may be determined by the difference of the normal components T_{11} and T_{33} of $\mathbf{t}$ and $\bar{\mathbf{t}}$.

Next, we use the representations (23.23) and (23.24) to analyze certain inhomogeneous motions, called *steady viscometric flows*. These motions are characterized by the condition that the deformation history at each point is of the form (23.8), but the shearing rate k may depend on the point.

23.1. Channel Flow

We assume (see Figure 2) that the fluid is confined between two fixed planes $x^1 = d$ and $x^1 = -d$, and we consider a flow with velocity field of the form

$$v^1 = 0, \qquad v^2 = v(x^1), \qquad v^3 = 0, \tag{23.29}$$

which satisfies the boundary condition

$$v(d) = v(-d) = 0. \tag{23.30}$$

Integrating the velocity field (23.29), we obtain the deformation functions

$$x^1 = X^1, \qquad x^2 = v(X^1)t + X^2, \qquad x^3 = X^3, \tag{23.31}$$

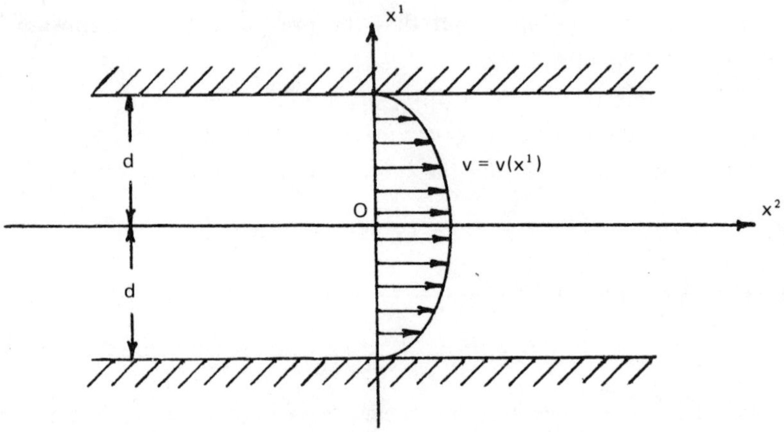

Figure 2.

where the configuration at time $t = 0$ is regarded as the reference configuration. From (23.31) we see that the motion is accelerationless, and that the relative deformation history is

$$[\mathbf{F}_t(t-s)] = \begin{bmatrix} 1 & 0 & 0 \\ -v'(x^1)s & 1 & 0 \\ 0 & 0 & 1 \end{bmatrix}. \tag{23.32}$$

This component matrix means that the motion is a simple shearing flow at each point with shearing rate

$$k = v'(x^1), \tag{23.33}$$

which depends on the coordinate x^1. Thus the motion is a steady viscometric flow. As a result, the representation (23.13) can be applied, and we have

$$T_{11} - T_{33} = \sigma_1(v'(x^1)), \tag{23.34a}$$

$$T_{22} - T_{33} = \sigma_2(v'(x^1)), \tag{23.34b}$$

$$T_{12} = \tau(v'(x^1)), \tag{23.34c}$$

which are all functions of x^1.

Note. Although the representation (23.13) was obtained for a simple shearing flow, it may be applied to a viscometric flow because for a simple fluid the stress is determined by the motion through the deformation history only. Since the history (23.32) coincides with the history (23.8) with k given by (23.33), the formulas (23.34) follow from (23.23) and (23.24).

Now under the assumption that the body force field vanishes, the linear momentum equation reduces to

$$\frac{\partial p}{\partial x^1} = \frac{dS_{11}}{dx^1}, \qquad \frac{\partial p}{\partial x^2} = \frac{dS_{12}}{dx^1}, \qquad \frac{\partial p}{\partial x^3} = 0. \tag{23.35}$$

Since the right-hand side of (23.35) depends only on x^1, the system (23.35) is integrable if and only if the pressure gradient in the direction of the channel is a constant, say,

$$\frac{\partial p}{\partial x^2} = \frac{dS_{12}}{dx^1} = -c. \tag{23.36}$$

Then the system (23.35) may be integrated, and the pressure field is

$$p = p(x^i, t) = -cx^2 + f(x^1) + g(t), \tag{23.37}$$

where f is a certain function of x^1 depending on the shearing rate $v'(x^1)$, while g is an arbitrary function of t.

The velocity profile $v(x^1)$ may be determined in the following way: First, since the boundary condition (23.30) is symmetric with respect to x^1, we may assume that v is an even function; i.e.,

$$v(x^1) = v(-x^1) \tag{23.38}$$

for all $x^1 \in [-d, d]$. Then $v'(x^1)$ is an odd function, and thus by virtue of (23.34c) and (23.15), $S_{12}(x^1)$ is also an odd function. We can determine the shear stress by integrating the differential equation (23.36), obtaining

$$S_{12}(x^1) = -cx^1, \tag{23.39}$$

where we have not added any integration constant because we know that $S_{12}(x^1)$ is odd with respect to x^1. Combining (23.39) with (23.34c), we get

$$\tau\big(v'(x^1)\big) = -cx^1. \tag{23.40}$$

Since τ is a monotonic function, this equation may be solved for $v'(x^1)$:

$$v'(x^1) = -\zeta(cx^1), \tag{23.41}$$

where ζ denotes the inverse function of τ. Integrating (23.41) and using the boundary condition (23.30), we finally obtain the velocity profile:

$$v(x^1) = \int_{x^1}^{d} \zeta(c\xi)\, d\xi. \tag{23.42}$$

Notice that the form of the function $v(x^1)$ depends explicitly on the response functional of the fluid through the function ζ. Thus each motion of the form (23.31) is a universal solution for all incompressible simple fluids having one particular shear stress function τ only. In other words, it is not possible to maintain the motion (23.31) with a particular velocity profile $v(x^1)$ by boundary forces alone for all incompressible simple fluids.

We can use the formula (23.42) for the velocity profile to test the shear stress function in the following way: Physically, the volume discharge rate through the channel is a convenient quantity to measure. Let Q be the discharge rate per unit channel width. Then

$$Q = Q(c) = \int_{-d}^{d} v(x^1)\, dx^1 = \int_{-d}^{d}\int_{x^1}^{d} \zeta(c\xi)\, d\xi\, dx^1 = \frac{2}{c}\int_{0}^{cd} \xi\zeta(\xi)\, d\xi, \tag{23.43}$$

where we have used integration by parts to reduce the double integral. The integral (23.43) is the solution of the differential equation

$$\zeta(cd) = \frac{1}{2cd^2} \frac{d}{dc} \left(c^2 Q(c)\right), \tag{23.44}$$

which determines the function ζ. Then the shear stress function τ may be obtained by inverting ζ.

Having considered the shear stress field in the motion, we turn our attention next to the normal stress fields. First, since the pressure field has the form (23.37), we have

$$T_{11}(x^i, t) = cx^2 - f(x^1) + g(t) + S_{11}(x^1). \tag{23.45}$$

But from (23.35a) the gradient of $T_{11}(x^i, t)$ in the direction of x^1 vanishes:

$$\frac{\partial T_{11}}{\partial x^1} = 0, \tag{23.46}$$

so we can rewrite (23.45) as

$$T_{11}(x^i, t) = cx^2 - g(t). \tag{23.47}$$

Then from (23.34) the other normal stress components are

$$\begin{aligned} T_{22}(x^i, t) &= cx^2 - g(t) + \sigma_2(v'(x^1)) - \sigma_1(v'(x^1)), \\ T_{33}(x^i, t) &= cx^2 - g(t) - \sigma_1(v'(x^1)). \end{aligned} \tag{23.48}$$

Also, the pressure field is

$$p(x^i, t) = -\tfrac{1}{3}\,\mathrm{tr}(\mathbf{T}(x^i, t)) = -cx^2 + g(t) + \tfrac{1}{3}[2\sigma_1(v'(x^1)) - \sigma_2(v'(x^1))]. \tag{23.49}$$

The formulas (23.40) and (23.46)–(23.49) determine completely the stress tensor field in the channel flow.

23.2. Poiseuille Flow

For this motion the fluid is confined in a circular pipe. We assume that the velocity field is of the form

$$v_{\langle r\rangle} = 0, \qquad v_{\langle\theta\rangle} = 0, \qquad v_{\langle z\rangle} = v(r) \tag{23.50}$$

relative to the cylindrical coordinate system (r, θ, z). The boundary condition is

$$v(R_0) = 0 \tag{23.51}$$

Integrating (23.50), we obtain the deformation functions

$$r = R, \qquad \theta = \Theta, \qquad z = v(R)t + Z, \tag{23.52}$$

where the reference configuration is the configuration at time $t = 0$, as before. From (23.52) we can prove easily that the motion is accelerationless, and that the physical component matrix of the relative deformation history is

$$[\mathbf{F}_t(t - s)] = \begin{bmatrix} 1 & 0 & 0 \\ 0 & 1 & 0 \\ -v'(r)s & 0 & 1 \end{bmatrix}. \tag{23.53}$$

Thus the motion is a viscometric flow with shearing rate

$$k = v'(r). \tag{23.54}$$

Hence the components of the stress tensor field satisfy the relations

$$T_{\langle rr\rangle} - T_{\langle\theta\theta\rangle} = \sigma_1(v'(r)), \tag{23.55a}$$

$$T_{\langle zz\rangle} - T_{\langle\theta\theta\rangle} = \sigma_2(v'(r)), \tag{23.55b}$$

$$T_{\langle rz\rangle} = \tau(v'(r)). \tag{23.55c}$$

Since the coordinate system (r, θ, z) is not a Cartesian system, we use the general formula (47.39) in Section 47, IVT-2, to calculate the components of div **T**. Specifically, the physical components of div **T** are given by

$$\begin{aligned}
(\operatorname{div}\mathbf{T})_{\langle r\rangle} &= \frac{\partial T_{\langle rr\rangle}}{\partial r} + \frac{1}{r}\frac{\partial T_{\langle r\theta\rangle}}{\partial\theta} + \frac{\partial T_{\langle rz\rangle}}{\partial z} + \frac{T_{\langle rr\rangle} - T_{\langle\theta\theta\rangle}}{r}, \\
(\operatorname{div}\mathbf{T})_{\langle\theta\rangle} &= \frac{\partial T_{\langle r\theta\rangle}}{\partial r} + \frac{1}{r}\frac{\partial T_{\langle\theta\theta\rangle}}{\partial\theta} + \frac{\partial T_{\langle\theta z\rangle}}{\partial z} + \frac{2}{r}T_{\langle r\theta\rangle}, \\
(\operatorname{div}\mathbf{T})_{\langle z\rangle} &= \frac{\partial T_{\langle rz\rangle}}{\partial r} + \frac{1}{r}\frac{\partial T_{\langle\theta z\rangle}}{\partial\theta} + \frac{\partial T_{\langle zz\rangle}}{\partial z} + \frac{1}{r}T_{\langle rz\rangle},
\end{aligned} \tag{23.56}$$

where we have also used the transformation rule (46.19) in Section 46, IVT-2, to express the results in physical components. Using (23.55) and (23.56), we see that the linear momentum equation reduces to

$$\frac{\partial p}{\partial r} = \frac{\partial S_{\langle rr\rangle}}{\partial r} + \frac{T_{\langle rr\rangle} - T_{\langle\theta\theta\rangle}}{r}, \qquad \frac{\partial p}{\partial\theta} = 0, \qquad \frac{\partial p}{\partial z} = \frac{\partial T_{\langle rz\rangle}}{\partial r} + \frac{T_{\langle rz\rangle}}{r}, \tag{23.57}$$

where we have assumed that the body force vanishes. Since the right-hand side of (23.57) depends only on r, the system is integrable if and only if the pressure gradient in the direction of the pipe is a constant, viz.,

$$\frac{\partial p}{\partial z} = \frac{\partial T_{\langle rz\rangle}}{\partial r} + \frac{T_{\langle rz\rangle}}{r} = -c. \tag{23.58}$$

Then the pressure field is of the form

$$p = -cz + f(r) + g(t). \tag{23.59}$$

The velocity profile $v(r)$ may be determined in a similar way as before. First, integrating (23.58) and using the fact that $T_{\langle rz\rangle}$ vanishes at $r = 0$ since the shearing rate vanishes there, we obtain

$$T_{\langle rz\rangle} = -\tfrac{1}{2}cr. \tag{23.60}$$

Hence from (23.55c) we get

$$v'(r) = -\zeta\left(\frac{cr}{2}\right). \tag{23.61}$$

Integrating this formula with respect to r and using the boundary condition (23.51), we obtain

$$v(r) = \int_r^{R_0} \zeta\left(\frac{c\xi}{2}\right) d\xi. \tag{23.62}$$

As before we can use the discharge rate Q to test the shear stress function. From (23.62)

$$\begin{aligned} Q = Q(c) &= \int_0^R 2\pi r v(r)\, dr = \int_0^{R_0} \int_r^{R_0} 2\pi r \zeta\left(\frac{c\xi}{2}\right) d\xi\, dr \\ &= \pi \int_0^{R_0} \xi^2 \zeta\left(\frac{c\xi}{2}\right) d\xi, \end{aligned} \tag{23.63}$$

which is the solution of the differential equation

$$\zeta\left(\frac{cR_0}{2}\right) = \frac{1}{\pi c^2 R_0^3} \frac{d}{dc}\left(c^3 Q(c)\right). \tag{23.64}$$

Inverting the function ζ, we obtain the shear stress function τ.

The normal stress components can be obtained from (23.55) as before.

The results are

$$\begin{aligned}
T_{\langle rr\rangle} &= cz - g(t) - \int_0^r \frac{1}{\xi}\,\sigma_1(v'(\xi))\,d\xi,\\
T_{\langle zz\rangle} &= cz - g(t) - \int_0^r \frac{1}{\xi}\,\sigma_1(v'(\xi))\,d\xi + \sigma_2(v'(r)) - \sigma_1(v'(r)),\\
T_{\langle\theta\theta\rangle} &= cz - g(t) - \int_0^r \frac{1}{\xi}\,\sigma_1(v'(\xi))\,d\xi - \sigma_1(v'(r)),\\
p &= -cz + g(t) + \int_0^r \frac{1}{\xi}\,\sigma_1(v'(\xi))\,d\xi + \frac{1}{3}\left[2\sigma_1(v'(r)) - \sigma_2(v'(r))\right].
\end{aligned} \tag{23.65}$$

23.3. Couette Flow

The fluid is confined between two concentric circular pipes, and the velocity field is of the form

$$v_{\langle r\rangle} = 0, \qquad v_{\langle\theta\rangle} = r\omega(r), \qquad v_{\langle z\rangle} = 0. \tag{23.66}$$

The boundary conditions are

$$\omega(R_1) = \Omega_1, \qquad \omega(R_2) = \Omega_2. \tag{23.67}$$

From (23.66) the deformation functions relative to the configuration at $t = 0$ are

$$r = R, \qquad \theta = \omega(R)t + \Theta, \qquad z = Z. \tag{23.68}$$

Unlike the previous two examples, the flow (23.68) has a centrifugal acceleration

$$a_{\langle r\rangle} = -r\omega(r)^2, \qquad a_{\langle\theta\rangle} = 0, \qquad a_{\langle z\rangle} = 0. \tag{23.69}$$

In physical components relative to (r, θ, z) the relative deformation history is

$$[\mathbf{F}_t(t-s)] = \begin{bmatrix} 1 & 0 & 0\\ -r\omega'(r)s & 1 & 0\\ 0 & 0 & 1 \end{bmatrix}. \tag{23.70}$$

Thus the motion is a viscometric flow with shearing rate

$$k = r\omega'(r), \tag{23.71}$$

which depends on r. Hence the components of the stress tensor field satisfy the relations

$$\begin{aligned} T_{\langle rr\rangle} - T_{\langle zz\rangle} &= \sigma_1(r\omega'(r)), \\ T_{\langle\theta\theta\rangle} - T_{\langle zz\rangle} &= \sigma_2(r\omega'(r)), \\ T_{\langle r\theta\rangle} &= \tau(r\omega'(r)). \end{aligned} \tag{23.72}$$

As before we assume that the body force vanishes. Then using (23.56), (23.69), and (23.72) we can express the linear momentum equation as

$$\frac{\partial p}{\partial r} = \frac{\partial S_{\langle rr\rangle}}{\partial r} + \frac{T_{\langle rr\rangle} - T_{\langle\theta\theta\rangle}}{r} + \varrho r\omega(r)^2, \tag{23.73a}$$

$$\frac{\partial p}{\partial \theta} = r\frac{\partial T_{\langle r\theta\rangle}}{\partial r} + 2T_{\langle r\theta\rangle}, \tag{23.73b}$$

$$\frac{\partial p}{\partial z} = 0. \tag{23.73c}$$

This system may be integrated directly; the pressure field is of the form

$$p = f(r) + g(t). \tag{23.74}$$

Note. The pressure gradient in the direction of the flow must vanish in this case, since $c\theta$ is not single valued unless $c = 0$.

The velocity profile may be determined in the following way: First, from (23.73b) with $\partial p/\partial\theta = 0$ we see that the shear stress is of the form

$$T_{\langle r\theta\rangle} = \frac{c}{r^2}, \tag{23.75}$$

where the constant of integration c may be regarded as being a twisting moment. Specifically, the torque per unit length in the z direction is given by

$$M = r2\pi rT_{\langle r\theta\rangle} = 2\pi r^2\frac{c}{r^2} = 2\pi c. \tag{23.76}$$

Substituting (23.75) and (23.76) into (23.72c), we get

$$r\omega'(r) = \zeta\left(\frac{M}{2\pi r^2}\right), \tag{23.77}$$

where ζ is the inverse function of τ as before. Integrating (23.77) with respect to r and using the boundary condition (23.67a), we obtain

$$\omega(r) = \Omega_1 + \int_{R_1}^{r}\frac{1}{\xi}\,\zeta\left(\frac{M}{2\pi\xi^2}\right)d\xi. \tag{23.78}$$

Now using the other boundary condition (23.67b), we have

$$\Delta\Omega = \Omega_2 - \Omega_1 = \int_{R_1}^{R_2} \frac{1}{\xi}\, \zeta\!\left(\frac{M}{2\pi\xi^2}\right) d\xi. \tag{23.79}$$

The formula (23.79) may be used to test the shear stress function. We define first

$$\varphi(M) = 2M\, \frac{d\,\Delta\Omega(M)}{dM}, \tag{23.80}$$

which may be obtained from (23.79) by Leibnitz's rule

$$\varphi(M) = \zeta\!\left(\frac{M}{2\pi R_1{}^2}\right) - \zeta\!\left(\frac{M}{2\pi R_2{}^2}\right). \tag{23.81}$$

Since $\varphi(M) \to 0$ as $M \to 0$, we can rewrite (23.81) as

$$\zeta\!\left(\frac{M}{2\pi R_1{}^2}\right) = \sum_{n=0}^{\infty} \varphi\!\left(M\left(\frac{R_1}{R_2}\right)^{2n}\right). \tag{23.82}$$

In the limiting case when R_2 is much greater than R_1, this formula reduces approximately to

$$\zeta\!\left(\frac{M}{2\pi R_1{}^2}\right) = \varphi(M) = 2M\, \frac{d\,\Delta\Omega(M)}{dM}. \tag{23.83}$$

In another limiting case when R_2 is very close to R_1, we can evaluate the integral (23.79) approximately by

$$\Delta\Omega(M) = \frac{\Delta R}{R_1}\, \zeta\!\left(\frac{M}{2\pi R_1{}^2}\right), \tag{23.84}$$

where ΔR denotes the difference of R_2 and R_1. Then the shear stress function is given by

$$\zeta\!\left(\frac{M}{2\pi R_1{}^2}\right) = R_1\, \frac{\Delta\Omega(M)}{\Delta R}. \tag{23.85}$$

The normal stress components may be obtained from (23.72) as before. The results are

$$T_{\langle rr\rangle} = T_{\langle rr\rangle}\,|_{R_1} + \int_{R_1}^{r} [\sigma_2(\xi\omega'(\xi)) - \sigma_1(\xi\omega'(\xi))]\, \frac{d\xi}{\xi} - \int_{R_1}^{r} \varrho\xi\omega(\xi)^2\, d\xi, \tag{23.86a}$$

$$T_{\langle\theta\theta\rangle} = T_{\langle rr\rangle} - \sigma_1(r\omega'(r)), \tag{23.86b}$$

$$T_{\langle zz\rangle} = T_{\langle rr\rangle} - \sigma_1(r\omega'(r)) + \sigma_2(r\omega'(r)), \tag{23.86c}$$

$$p = T_{\langle rr\rangle} + \tfrac{1}{3}[2\sigma_1(r\omega'(r)) - \sigma_2(r\omega'(r))]. \tag{23.86d}$$

The formula (23.86a) may be used to test the difference of the normal stress functions. Evaluating (23.86a) at $r = R_2$, we get

$$\Delta T_{\langle rr\rangle} = T_{\langle rr\rangle}\,|_{R_2} - T_{\langle rr\rangle}\,|_{R_1}$$
$$= \int_{R_1}^{R_2} \left\{ \frac{1}{\xi} \left[\sigma_2(\xi\omega'(\xi)) - \sigma_1(\xi\omega'(\xi)) \right] - \varrho\xi\omega(\xi)^2 \right\} d\xi. \qquad (23.87)$$

We define the function

$$\psi(M) \equiv 2M \frac{d}{dM}\left(\Delta T_{\langle rr\rangle} + \int_{R_1}^{R_2} \varrho\xi\omega(\xi)^2\, d\xi \right). \qquad (23.88)$$

Then from (23.87)

$$\psi(M) = [\sigma_2 - \sigma_1]\left(\zeta\left(\frac{M}{2\pi R_1^2}\right)\right) - [\sigma_2 - \sigma_1]\left(\zeta\left(\frac{M}{2\pi R_2^2}\right)\right), \qquad (23.89)$$

which implies that

$$[\sigma_2 - \sigma_1]\left(\zeta\left(\frac{M}{2\pi R_1^2}\right)\right) = \sum_{n=0}^{\infty} \psi\left(M\left(\frac{R_1}{R_2}\right)^{2n}\right). \qquad (23.90)$$

In particular, when R_2 is much greater than R_1, we have the approximate formula

$$[\sigma_2 - \sigma_1]\left(\zeta\left(\frac{M}{2\pi R_1^2}\right)\right) = \psi(M). \qquad (23.91)$$

On the other hand, when R_2 is very close to R_1, we obtain directly from (23.87) the approximate formula

$$[\sigma_2 - \sigma_1]\left(\zeta\left(\frac{M}{2\pi R_1^2}\right)\right) = \varrho R_1^2 \Omega_1 \Omega_2 + R_1 \frac{\Delta T_{\langle rr\rangle}}{\Delta R}. \qquad (23.92)$$

Of course, these formulas do not determine the normal stress functions σ_2 and σ_1 individually. However, there are other viscometric flows which allow us to test one of the two normal stress functions. Then (23.91) or (23.92) determine uniquely the other one.

For more details on the viscometric flows of incompressible simple fluids we refer the reader to the encyclopedia article by Truesdell and Noll.[(6)]

[(6)] C. Truesdell and W. Noll, The non-linear field theories of mechanics, in Flügge's *Handbuch der Physik*, Vol. III/3, Springer-Verlag, Berlin, 1965.

24. Universal Solutions for Isotropic Elastic Solids I: The Compressible Case

In the preceding section we have analyzed some flows for the class of incompressible simple fluids. Now we turn our attention to the response of the class of isotropic elastic solids. Mathematically, this particular class of materials is convenient since there is a simple representation formula (19.42) for the constitutive equation. In application this class is also quite important. In fact the linearization of (19.42) at a stress-free natural state is the basis of a major part of the classical theory of linear elasticity.

Since the material is assumed to be elastic, its response in static deformations determines the constitutive equation completely. Hence the problem of finding equilibrium configurations is just as important as the problem of determining motions. In Section 22 we introduced the concept of a universal solution for the dynamic problem. Clearly, this concept may be applied to the static problem also. Specifically, let $\boldsymbol{\varkappa}$ be a homogeneous reference configuration of a body manifold $\mathscr{B}$ as before. Then a (static) deformation $\boldsymbol{\varphi}$ relative to $\boldsymbol{\varkappa}$ is called a *static universal solution* for a class of materials if the deformed configuration $\boldsymbol{\varphi}(\boldsymbol{\varkappa}(\mathscr{B})) = \boldsymbol{\chi}(\mathscr{B})$ may be maintained in equilibrium by boundary forces alone regardless of what material belonging to the class the body $\mathscr{B}$ is made up of. We now consider the problem of static universal solutions for the class of isotropic elastic solids.

In Section 22 we pointed out that an accelerationless homogeneous motion is a universal solution for all simple materials. In particular, when the motion is time independent, the deformation becomes a static universal solution. We now show that the homogeneous deformations are the only static universal solutions for any type of elastic solids. This result is a consequence of Ericksen's theorem, which asserts that any static universal solution of the class of isotropic hyperelastic solids must be a homogeneous deformation. As a result, any static universal solution of an arbitrary type of elastic or hyperelastic solids with a particular symmetry group must also be a homogeneous deformation, since the symmetry group of the isotropic solids contains the symmetry group of any other type of solids.

We recall that in Section 20 we defined a hyperelastic material by a constitutive equation of the form

$$\mathbf{T} = \varrho \mathbf{F}\left(\frac{\partial \varepsilon}{\partial \mathbf{F}}\right)^T, \tag{24.1}$$

where $\varepsilon = \varepsilon(\mathbf{F})$ denotes the stored energy function. Suppose that the hyperelastic material is an isotropic solid, and let $\varkappa$ be an undistorted reference configuration. Then ε satisfies the conditions

$$\varepsilon(\mathbf{QF}) = \varepsilon(\mathbf{F}) \tag{24.2}$$

and

$$\varepsilon(\mathbf{FQ}) = \varepsilon(\mathbf{F}) \tag{24.3}$$

for all rotations $\mathbf{Q}$. A representation for these conditions may be found easily by using an argument similar to that in the proof of (19.42):

$$\varepsilon(\mathbf{F}) = f(\mathrm{I}_{\mathbf{V}}, \mathrm{II}_{\mathbf{V}}, \mathrm{III}_{\mathbf{V}}), \tag{24.4}$$

where $\mathrm{I}_{\mathbf{V}}, \mathrm{II}_{\mathbf{V}}, \mathrm{III}_{\mathbf{V}}$ denote the three fundamental invariants of the left stretch tensor $\mathbf{V}$ of $\mathbf{F}$. Indeed, it follows from (24.3) that ε depends on $\mathbf{F}$ through $\mathbf{V}$. Then from (24.2) we see that ε is an isotropic function of $\mathbf{V}$. Thus the representation is of the form (24.4).

In application it is more convenient to use the variable $\mathbf{B}$ defined by

$$\mathbf{B} = \mathbf{V}^2 = \mathbf{F}\mathbf{F}^T \tag{24.5}$$

instead of the variable $\mathbf{V}$, since $\mathbf{B}$ is a polynomial of $\mathbf{F}$ while $\mathbf{V}$ is not. Notice that both $\mathbf{B}$ and $\mathbf{V}$ are positive definite and symmetric, so they determine each other uniquely. Thus a function of $\mathbf{V}$ may be regarded as a function of $\mathbf{B}$, and vice versa. We call $\mathbf{B}$ the *left Cauchy–Green tensor* of $\mathbf{F}$. Using this tensor, we can rewrite the representation (24.4) as

$$\varepsilon(\mathbf{F}) = g(\mathrm{I}_{\mathbf{B}}, \mathrm{II}_{\mathbf{B}}, \mathrm{III}_{\mathbf{B}}). \tag{24.6}$$

Substituting (24.5) and (24.6) into (24.1) and using the chain rule, we obtain

$$\mathbf{T} = 2\varrho\mathbf{B}\left(\frac{\partial g}{\partial \mathrm{I}_{\mathbf{B}}}\,\frac{\partial \mathrm{I}_{\mathbf{B}}}{\partial \mathbf{B}} + \frac{\partial g}{\partial \mathrm{II}_{\mathbf{B}}}\,\frac{\partial \mathrm{II}_{\mathbf{B}}}{\partial \mathbf{B}} + \frac{\partial g}{\partial \mathrm{III}_{\mathbf{B}}}\,\frac{\partial \mathrm{III}_{\mathbf{B}}}{\partial \mathbf{B}}\right). \tag{24.7}$$

Now the partial derivatives of the invariants $\mathrm{I}_{\mathbf{B}}, \mathrm{II}_{\mathbf{B}}, \mathrm{III}_{\mathbf{B}}$ with respect to $\mathbf{B}$ may be calculated in the following way: We derive first the formula for the partial derivative of the determinant of an arbitrary invertible tensor $\mathbf{A}$ with respect to $\mathbf{A}$, viz.,

$$\frac{\partial \det \mathbf{A}}{\partial \mathbf{A}} = (\det \mathbf{A})(\mathbf{A}^{-1})^T. \tag{24.8}$$

To prove this basic formula we write

$$\det(\mathbf{A} + \lambda\mathbf{C}) = \lambda^3(\det \mathbf{A}) \det\left(\frac{1}{\lambda}\mathbf{I} + \mathbf{A}^{-1}\mathbf{C}\right). \tag{24.9}$$

Now using the characteristic polynomial for the tensor $\mathbf{A}^{-1}\mathbf{C}$ (cf. Section 26, IVT-1), we can expand the right-hand side of (24.9) as

$$\det(\mathbf{A} + \lambda\mathbf{C}) = (\det \mathbf{A})[1 + \mathrm{I}_{\mathbf{A}^{-1}\mathbf{C}}\lambda + \mathrm{II}_{\mathbf{A}^{-1}\mathbf{C}}\lambda^2 + \mathrm{III}_{\mathbf{A}^{-1}\mathbf{C}}\lambda^3]. \quad (24.10)$$

Differentiating this expression with respect to λ and evaluating the result at $\lambda = 0$, we get

$$\frac{d}{d\lambda}\det(\mathbf{A} + \lambda\mathbf{C})\,|_{\lambda=0} = (\det \mathbf{A})\mathrm{I}_{\mathbf{A}^{-1}\mathbf{C}} = \operatorname{tr}\big((\det \mathbf{A})\mathbf{A}^{-1}\mathbf{C}\big). \quad (24.11)$$

Since $\mathbf{C}$ is an arbitrary tensor, this result implies directly the formula (24.8). Since $\mathrm{III}_{\mathbf{B}}$ is just the determinant of $\mathbf{B}$,

$$\frac{\partial \mathrm{III}_{\mathbf{B}}}{\partial \mathbf{B}} = \mathrm{III}_{\mathbf{B}}\mathbf{B}^{-1}. \quad (24.12)$$

The partial derivative of the other two invariants may now be derived as follows: Applying (24.12) to the invertible tensor $\lambda\mathbf{I} + \mathbf{B}$ for sufficiently small λ, we have

$$\frac{\partial}{\partial \mathbf{B}}\det(\lambda\mathbf{I} + \mathbf{B}) = \det(\lambda\mathbf{I} + \mathbf{B})(\lambda\mathbf{I} + \mathbf{B})^{-1}. \quad (24.13)$$

Since $\det(\lambda\mathbf{I} + \mathbf{B})$ may be expressed by the characteristic polynomial, we have

$$\frac{\partial}{\partial \mathbf{B}}\det(\lambda\mathbf{I} + \mathbf{B}) = \frac{\partial \mathrm{I}_{\mathbf{B}}}{\partial \mathbf{B}}\lambda^2 + \frac{\partial \mathrm{II}_{\mathbf{B}}}{\partial \mathbf{B}}\lambda + \frac{\partial \mathrm{III}_{\mathbf{B}}}{\partial \mathbf{B}}. \quad (24.14)$$

Thus (24.13) implies that

$$\left(\mathbf{I}\frac{\partial \mathrm{I}_{\mathbf{B}}}{\partial \mathbf{B}}\right)\lambda^3 + \left(\mathbf{B}\frac{\partial \mathrm{I}_{\mathbf{B}}}{\partial \mathbf{B}} + \mathbf{I}\frac{\partial \mathrm{II}_{\mathbf{B}}}{\partial \mathbf{B}}\right)\lambda^2 + \left(\mathbf{B}\frac{\partial \mathrm{II}_{\mathbf{B}}}{\partial \mathbf{B}} + \mathbf{I}\frac{\partial \mathrm{III}_{\mathbf{B}}}{\partial \mathbf{B}}\right)\lambda + \mathbf{B}\frac{\partial \mathrm{III}_{\mathbf{B}}}{\partial \mathbf{B}}$$
$$= \mathbf{I}\lambda^3 + \mathrm{I}_{\mathbf{B}}\mathbf{I}\lambda^2 + \mathrm{II}_{\mathbf{B}}\mathbf{I}\lambda + \mathrm{III}_{\mathbf{B}}\mathbf{I}. \quad (24.15)$$

Matching the coefficients of λ^3 and λ^2, we get

$$\frac{\partial \mathrm{I}_{\mathbf{B}}}{\partial \mathbf{B}} = \mathbf{I}, \quad (24.16\text{a})$$

$$\frac{\partial \mathrm{II}_{\mathbf{B}}}{\partial \mathbf{B}} = \mathrm{I}_{\mathbf{B}}\mathbf{I} - \mathbf{B}. \quad (24.16\text{b})$$

Note. We can recover also the formula (24.12) and the Cayley–Hamilton equation for $\mathbf{B}$ [cf. equation (26.14) in Section 26, IVT-1] by matching the coefficients of λ^1 and λ^0 in (24.15).

Substituting (24.12) and (24.16) into (24.7), we obtain a representation formula for the constitutive equation of an isotropic hyperelastic solid:

$$\mathbf{T} = 2\varrho \frac{\partial g}{\partial \mathrm{III}_{\mathbf{B}}} \mathrm{III}_{\mathbf{B}}\mathbf{I} + 2\varrho\left(\frac{\partial g}{\partial \mathrm{I}_{\mathbf{B}}} + \frac{\partial g}{\partial \mathrm{II}_{\mathbf{B}}}\right)\mathbf{B} - 2\varrho \frac{\partial g}{\partial \mathrm{II}_{\mathbf{B}}} \mathbf{B}^2. \tag{24.17}$$

Notice that the coefficients of $\mathbf{I}$, $\mathbf{B}$, and $\mathbf{B}^2$ in the formula are functions of the fundamental invariants $\mathrm{I}_{\mathbf{B}}, \mathrm{II}_{\mathbf{B}}, \mathrm{III}_{\mathbf{B}}$. Thus (24.17) is a special case of the general formula (19.42) when the variable $\mathbf{V}$ is replaced by the variable $\mathbf{B}$. Since the coefficients in (24.17) are derived from a single stored energy function g, they must satisfy certain compatibility conditions, as we have remarked in Section 20.

Since $\mathbf{B}$ and $\mathbf{B}^{-1}$ determine each other uniquely, and since their sets of fundamental invariants also determine each other uniquely, we can express the stored energy function ε as a function of the invariants $\mathrm{I}_1, \mathrm{I}_2, \mathrm{I}_3$ of $\mathbf{B}^{-1}$, viz.,

$$\varepsilon(\mathbf{F}) = h(\mathrm{I}_1, \mathrm{I}_2, \mathrm{I}_3). \tag{24.18}$$

In the proof of Ericksen's theorem it is more convenient to use the representation (24.18) than the earlier one, (24.6), because the components of $\mathbf{B}^{-1}$ are precisely the covariant components of the Euclidean metric tensor on $\boldsymbol{\varkappa}(\mathbf{B})$, when the coordinates are transformed from (X^A) to (x^i) by the deformation functions. That is,

$$(B^{-1})_{ij} = \delta_{AB} \frac{\partial X^A}{\partial x^i} \frac{\partial X^B}{\partial x^j}. \tag{24.19}$$

As a result, the curvature tensor based on $\mathbf{B}^{-1}$ must vanish (cf. Exercise 47.7 in Section 47, IVT-2, and the Theorem 59.1 in Section 59, IVT-2). This condition of integrability may be obtained by differentiating (24.19) with respect to x^k and then using the symmetry condition for the partial derivatives. The result is

$$\begin{aligned} &\tfrac{1}{2}(B^{-1}_{km,pq} + B^{-1}_{qp,km} - B^{-1}_{kp,mq} - B^{-1}_{qm,kp}) \\ &\qquad + B_{rs}(A_{qpr}A_{kms} - A_{qmr}A_{kps}) = 0, \end{aligned} \tag{24.20}$$

where the comma denotes partial derivative with respect to x^k, e.g., $B^{-1}_{km,pq} = \partial^2 B^{-1}_{km}/\partial x^p\, \partial x^q$, etc., and where

$$A_{kmp} = A_{mkp} = \tfrac{1}{2}(B^{-1}_{kp,m} + B^{-1}_{mp,k} - B^{-1}_{km,p}). \tag{24.21}$$

Now in terms of $\mathbf{B}^{-1}$ and its invariants I_a the constitutive equation becomes

$$\mathbf{T} = -2\varrho\mathbf{B}^{-1}\left(\frac{\partial h}{\partial \mathrm{I}_1}\frac{\partial \mathrm{I}_1}{\partial \mathbf{B}^{-1}} + \frac{\partial h}{\partial \mathrm{I}_2}\frac{\partial \mathrm{I}_2}{\partial \mathbf{B}^{-1}} + \frac{\partial h}{\partial \mathrm{I}_3}\frac{\partial \mathrm{I}_3}{\partial \mathbf{B}^{-1}}\right)$$
$$= -2\varrho\frac{\partial h}{\partial \mathrm{I}_3}\mathrm{I}_3\mathbf{I} - 2\varrho\left(\frac{\partial h}{\partial \mathrm{I}_1} + \frac{\partial h}{\partial \mathrm{I}_2}\mathrm{I}_1\right)\mathbf{B}^{-1} + 2\varrho\frac{\partial h}{\partial \mathrm{I}_2}\mathbf{B}^{-2}. \qquad (24.22)$$

We shall now use this representation to establish the following important result:

Theorem (Ericksen[(7)]). A deformation $\boldsymbol{\varphi}$ relative to a homogeneous reference configuration $\boldsymbol{\varkappa}$ is possible for every isotropic hyperelastic material subject to boundary forces alone if and only if it is homogeneous.

To prove this theorem we seek necessary conditions on the deformation from the equation of equilibrium

$$\operatorname{div}\mathbf{T} = \mathbf{0}, \qquad (24.23)$$

where the body force vanishes since the deformation is required to be a universal solution. Replacing the density ϱ in (24.22) by $\varrho_{\varkappa}\mathrm{I}_3{}^{1/2}$ and regarding $\varrho_{\varkappa}$ as a constant, we obtain from (24.22) and (24.23)

$$\frac{\partial h}{\partial \mathrm{I}_a}\frac{\partial}{\partial x^m}\left(\mathrm{I}_3{}^{1/2}B^{-1}_{kp}\frac{\partial \mathrm{I}_a}{\partial B^{-1}_{pm}}\right) + \mathrm{I}_3{}^{1/2}B^{-1}_{kp}\frac{\partial \mathrm{I}_a}{\partial B^{-1}_{pm}}\frac{\partial^2 h}{\partial \mathrm{I}_a\,\partial \mathrm{I}_b}\frac{\partial \mathrm{I}_b}{\partial x^m} = 0, \qquad (24.24)$$

where the repeated indices a and b are summed from 1 to 3. In order that the deformation be a universal solution, the field equation (24.24) must hold for all forms of the stored energy function h. Hence the coefficients of $\partial h/\partial \mathrm{I}_a$ must vanish, and the coefficients of $\partial^2 h/\partial \mathrm{I}_a\,\partial \mathrm{I}_b$ must be skew symmetric with respect to the indices a and b; i.e.,

$$\frac{\partial}{\partial x^m}\left(\mathrm{I}_3{}^{1/2}B^{-1}_{kp}\frac{\partial \mathrm{I}_a}{\partial B^{-1}_{pm}}\right) = 0, \qquad (24.25)$$

and

$$\left(\frac{\partial \mathrm{I}_a}{\partial B^{-1}_{pm}}\frac{\partial \mathrm{I}_b}{\partial x^m} + \frac{\partial \mathrm{I}_b}{\partial B^{-1}_{pm}}\frac{\partial \mathrm{I}_a}{\partial x^m}\right)B^{-1}_{kp} = 0, \qquad (24.26)$$

where a and b are now free indices.

(7) J. L. Ericksen, Deformation possible in every compressible, isotropic, perfectly elastic material, *Journal of Mathematical Physics*, Vol. 34, pp. 126–128, 1955. Reprinted in *Continuum Mechanics*, International Science Review Series, Vol. 8, Edited by C. Truesdell, Gordon and Breach, New York, 1965.

Choosing $a = 3$ in (24.25) and using (24.12) for $\mathbf{B}^{-1}$, we get

$$\frac{\partial}{\partial x^m}\left(\mathrm{I}_3^{\,1/2}\frac{\partial \mathrm{I}_3}{\partial B^{-1}_{pm}}B^{-1}_{kp}\right) = \frac{\partial}{\partial x^m}(\mathrm{I}_3^{\,3/2}B_{mp}B^{-1}_{kp}) = \frac{\partial}{\partial x^m}(\mathrm{I}_3^{\,3/2}) = 0, \tag{24.27}$$

so I_3 is necessarily a constant field. Next, choosing $a = 1$ in (24.25) and using (24.16a) for $\mathbf{B}^{-1}$, we get

$$\mathrm{I}_3^{\,1/2}\frac{\partial}{\partial x^m}\left(\frac{\partial \mathrm{I}_1}{\partial B^{-1}_{pm}}B^{-1}_{kp}\right) = \mathrm{I}_3^{\,1/2}\frac{\partial}{\partial x^m}(\delta_{pm}B^{-1}_{kp}) = \mathrm{I}_3^{\,1/2}B^{-1}_{km,m} = 0. \tag{24.28}$$

Also, choosing $a = 3$ and $b = 1$ in (24.26), we have

$$\frac{\partial \mathrm{I}_1}{\partial x^m}\frac{\partial \mathrm{I}_3}{\partial B^{-1}_{pm}}B^{-1}_{kp} = \frac{\partial \mathrm{I}_1}{\partial x^m}\mathrm{I}_3B_{pm}B^{-1}_{kp} = \mathrm{I}_3\frac{\partial \mathrm{I}_1}{\partial x^m} = \mathrm{I}_3B^{-1}_{kk,m} = 0, \tag{24.29}$$

so I_1 is necessarily a constant field. Now substituting the results (24.28) and (24.29) into (24.21) and setting $k = m$, we get

$$A_{kkp} = \tfrac{1}{2}(B^{-1}_{kp,k} + B^{-1}_{kp,k} - B^{-1}_{kk,p}) = 0. \tag{24.30}$$

Next, using the results (24.28)–(24.30) and setting $p = q$ and $k = m$ in the integrability condition (24.20), we obtain

$$B_{rs}A_{pkr}A_{kps} = (V_{rq}A_{pkr})(V_{sq}A_{kps}) = 0, \tag{24.31}$$

where $\mathbf{V}$ is the left stretch tensor which is also the square root of $\mathbf{B}$; cf. (24.5). Since (24.31) reduces to a sum of squares, we have

$$V_{rq}A_{pkr} = 0, \tag{24.32}$$

which implies that

$$A_{pkr} = 0. \tag{24.33}$$

Then from (24.21)

$$B^{-1}_{ps,k} = A_{kps} + A_{skp} = 0. \tag{24.34}$$

Thus $\mathbf{B}^{-1}$ is necessarily a constant field.

By virtue of (24.19) we know that B^{-1}_{ij} corresponds to the covariant components of the Euclidean metric on $\boldsymbol{\varkappa}(\mathscr{B})$ expressed in the coordinate system (x^i). Now the condition (24.34) implies that the Christoffel symbols of that metric vanish in both (X^A) and (x^i). In Section 47, IVT-2, we have derived a general transformation law for the Christoffel symbols. In particular, when the Christoffel symbols vanish in both (X^A) and (x^i), the transformation law reduces to

$$0 = \frac{\partial x^k}{\partial X^A} \frac{\partial^2 X^A}{\partial x^i \, \partial x^j} + 0, \tag{24.35}$$

which implies directly that

$$\frac{\partial^2 X^A}{\partial x^i \, \partial x^j} = F^{-1}_{Ai,j} = 0. \tag{24.36}$$

Thus **F** is necessarily a constant field; i.e., the deformation is homogeneous, and the proof is complete.

Since a hyperelastic material is *ipso facto* an elastic material, Ericksen's theorem shows that a universal solution for all isotropic elastic solids must be a homogeneous deformation. Next, since the response function of an isotropic elastic material satisfies automatically the symmetry condition of any type of solids, the theorem implies also that a universal solution for any type of elastic solids must be a homogeneous deformation. The symmetry group of an elastic fluid, however, is bigger than the symmetry group of an isotropic solid. Indeed, the constitutive equation of an elastic fluid has the representation

$$\mathbf{T} = -p(\varrho)\mathbf{I}, \tag{24.37}$$

which is an extremely simple special case of (19.42). As a result, there are a lot more universal solutions for elastic fluids. In fact from (24.37) the equation of equilibrium reduces to

$$-p'(\varrho) \operatorname{grad} \varrho = 0. \tag{24.38}$$

Thus a deformation is a universal solution if and only if det **F** is a constant field. Obviously, this condition can be satisfied by all homogeneous deformations as well as by some inhomogeneous ones.

Having solved the problem of static universal solutions for all types of elastic solids, we consider next the problem of dynamic universal solutions. This problem has a very simple answer. Namely, a motion is a dynamic universal solution for any type of elastic solid if and only if it is an accelerationless homogeneous motion. Such a motion has been considered in Section 22.

The reason for the preceding result is more or less obvious. Since both the condition of symmetry and the condition of frame-indifference are linear with respect to the response function, a dynamic universal solution for any type of elastic material must be accelerationless, and the instantaneous deformation of the motion must be a static universal solution. Indeed, if a motion satisfies the linear momentum equation

$$\operatorname{div} \mathbf{G}(\mathbf{F}) = \varrho \mathbf{a} \tag{24.39}$$

for a collection of $\mathbf{G}$ which is closed with respect to linear operations, then we have also

$$\operatorname{div} c\mathbf{G}(\mathbf{F}) = \varrho \mathbf{a} \tag{24.40}$$

for any constant c. As a result, (24.39) reduces to

$$\operatorname{div} \mathbf{G}(\mathbf{F}) = 0, \qquad \mathbf{a} = \mathbf{0}. \tag{24.41}$$

Consequently, for any type of elastic materials, a dynamic universal solution is simply an accelerationless motion such that the instantaneous deformations are all static universal solutions. In particular, for any type of elastic solids, a dynamic universal solution is just an accelerationless homogeneous motion.

The preceding criterion for a dynamic universal solution is not applicable to incompressible materials. For such materials the acceleration field may be balanced by the gradient of a pressure field. We shall consider the problem of universal solutions for incompressible elastic solids in the following section.

25. Universal Solutions for Isotropic Elastic Solids II: The Incompressible Case

For an incompressible isotropic elastic solid the constitutive equation is

$$\mathbf{T} = -p\mathbf{I} + f_1\mathbf{B} + f_2\mathbf{B}^2, \tag{25.1}$$

where f_1 and f_2 are functions of $\mathrm{I}_{\mathbf{B}}$ and $\mathrm{II}_{\mathbf{B}}$. As before we assume that the reference configuration itself also satisfies the constraint. Then we have

$$\mathrm{III}_{\mathbf{B}} = \det \mathbf{B} = 1. \tag{25.2}$$

From (25.2) we can verify easily that

$$\mathrm{I}_{\mathbf{B}} = \mathrm{II}_{\mathbf{B}^{-1}} \equiv \mathrm{I}, \qquad \mathrm{II}_{\mathbf{B}} = \mathrm{I}_{\mathbf{B}^{-1}} \equiv \mathrm{II}. \tag{25.3}$$

Also, by use of the Cayley–Hamilton equation [cf. (26.14) in Section 26, IVT-1] we can rewrite (25.1) as

$$\mathbf{T} = -p\mathbf{I} + g_1\mathbf{B} + g_{-1}\mathbf{B}^{-1}, \tag{25.4}$$

where g_1 and g_{-1} are functions of I and II.

If the material is hyperelastic, then there is a stored energy function ε which has the representation

$$\varepsilon = h(\mathrm{I}, \mathrm{II}). \tag{25.5}$$

In this case the constitutive equation becomes

$$\mathbf{T} = -p\mathbf{I} + 2\varrho\left(\frac{\partial h}{\partial \mathrm{I}} + \frac{\partial h}{\partial \mathrm{II}}\,\mathrm{I}\right)\mathbf{B} - 2\varrho\,\frac{\partial h}{\partial \mathrm{II}}\,\mathbf{B}^2 \tag{25.6}$$

or, equivalently,

$$\mathbf{T} = -p\mathbf{I} + 2\varrho\,\frac{\partial h}{\partial \mathrm{I}}\,\mathbf{B} - 2\varrho\,\frac{\partial h}{\partial \mathrm{II}}\,\mathbf{B}^{-1}, \tag{25.7}$$

where ϱ is equal to the constant density ϱ_κ, since the deformation must satisfy the condition of constraint (25.2).

The problem of static universal solutions for incompressible isotropic elastic materials in general may be formulated as before by requiring the equation of equilibrium

$$\operatorname{div}(g_1\mathbf{B} + g_{-1}\mathbf{B}^{-1}) = \operatorname{grad} p \tag{25.8}$$

to be integrable for p for all choice of g_1 and g_2. Similarly, if the materials are hyperelastic, then the problem is defined by the equation

$$\operatorname{div}\left(\frac{\partial h}{\partial \mathrm{I}}\,\mathbf{B} - \frac{\partial h}{\partial \mathrm{II}}\,\mathbf{B}^{-1}\right) = \operatorname{grad} p \tag{25.9}$$

for all choice of h. The problem (25.9) was considered by Ericksen.[8] Unfortunately, a complete answer is not known. So far, we only know that the homogeneous isochoric deformations and five families of inhomogeneous deformations, which we shall summarize later in this section, come near to exhausting[9] all solutions. The determination of the complete

[8] J. L. Ericksen, Deformation possible in every isotropic, incompressible, perfectly elastic body, *Zeitschrift für Angewandte Mathematik und Physik*, Vol. 5, pp. 466–489, 1954. Reprinted in *Continuum Mechanics*, International Science Review Series, Vol. 8, edited by C. Truesdell, Gordon and Breach, New York, 1965.

[9] See, for example, A. W. Marris, and J. F. Shiau, Universal deformations in isotropic incompressible hyperelastic materials when the deformation tensor has equal proper

solutions, however, remains a major unsolved problem in the theory of elasticity.

Since there is no definitive answer to the problem of universal solutions for incompressible isotropic elastic solids, we shall not summarize here any of the partial analyses, which are all very lengthy. In the remainder of this section we shall consider only the five presently known families of inhomogeneous universal solutions. Our analysis follows closely the encyclopedia article by Truesdell and Noll[(10)] and the recent book by Wang and Truesdell.[(11)]

The deformation functions of the five known families of inhomogeneous universal solutions are given in terms of the Cartesian coordinate system, the cylindrical coordinate system, and the spherical coordinate system, the latter two being curvilinear systems. In Section 23 we have given the formulas for the physical components of div **T** relative to the cylindrical coordinate system; cf. (23.56). Now we give the same relative to the spherical coordinate system (r, θ, φ). The derivation is based on the formulas (46.19) and (47.39) in Sections 46 and 47, IVT-2, as before. The results are

$$
\begin{aligned}
(\operatorname{div} \mathbf{T})_{\langle r\rangle} &= \frac{\partial T_{\langle rr\rangle}}{\partial r} + \frac{1}{r}\frac{\partial T_{\langle r\theta\rangle}}{\partial \theta} + \frac{1}{r\sin\theta}\frac{\partial T_{\langle r\varphi\rangle}}{\partial \varphi} \\
&\quad + \frac{1}{r}\,[2T_{\langle rr\rangle} - T_{\langle\theta\theta\rangle} - T_{\langle\varphi\varphi\rangle} + \cot\theta\, T_{\langle r\theta\rangle}], \\
(\operatorname{div} \mathbf{T})_{\langle\theta\rangle} &= \frac{\partial T_{\langle r\theta\rangle}}{\partial r} + \frac{1}{r}\frac{\partial T_{\langle\theta\theta\rangle}}{\partial \theta} + \frac{1}{r\sin\theta}\frac{\partial T_{\langle\theta\varphi\rangle}}{\partial \varphi} \\
&\quad + \frac{1}{r}\,[3T_{\langle r\theta\rangle} + \cot\theta(T_{\langle\theta\theta\rangle} - T_{\langle\varphi\varphi\rangle})], \\
(\operatorname{div} \mathbf{T})_{\langle\varphi\rangle} &= \frac{\partial T_{\langle r\varphi\rangle}}{\partial r} + \frac{1}{r}\frac{\partial T_{\langle\theta\varphi\rangle}}{\partial \theta} + \frac{1}{r\sin\theta}\frac{\partial T_{\langle\varphi\varphi\rangle}}{\partial \varphi} \\
&\quad + \frac{1}{r}\,[3T_{\langle r\varphi\rangle} + 2\cot\theta\, T_{\langle\theta\varphi\rangle}].
\end{aligned}
\tag{25.10}
$$

values, *Archive for Rational Mechanics and Analysis*, Vol. 36, pp. 135–160, 1970; R. L. Fosdick and K. W. Schaler, On Ericksen's problem for plane deformations with uniform transverse stretch, *International Journal of Engineering Science*, Vol. 7, pp. 217–233, 1969; W. C. Müller, Some further results on the Ericksen-problem for deformations with constant strain invariants, *Zeitschrift für Angewandte Mathematik und Physik*, Vol. 21, pp. 633–636 (1970); C. B. Kafadar, On Ericksen's problem, *Archive for Rational Mechanics and Analysis*, Vol. 47, pp. 15–27, 1972.

(10) C. Truesdell and W. Noll, The non-linear field theories of mechanics, in Flügge's *Handbuch der Physik*, Vol. III/3, Springer-Verlag, Berlin, 1965.

(11) C.-C. Wang, and C. Truesdell, *Introduction to Rational Elasticity*, Noordhoff International Publishing Company, Leyden, 1973.

We consider first some special types of stress tensor fields. Direct calculation shows that these fields satisfy the equations of equilibrium without any body force field.

25.1. Planar Problems

We assume that the physical components of the determinate stress $\mathbf{S} = \mathbf{T} + p\mathbf{I}$ relative to a Cartesian system (x, y, z) satisfy the following conditions: (a) the shear components $S_{\langle xy\rangle}$ and $S_{\langle xz\rangle}$ vanish, and (b) the remaining nonzero components of $\mathbf{S}$ depend only on x. Under these conditions the equations of equilibrium reduce to

$$\frac{\partial p}{\partial x} = \frac{dS_{\langle xx\rangle}}{dx}, \qquad \frac{\partial p}{\partial y} = 0, \qquad \frac{\partial p}{\partial z} = 0. \tag{25.11}$$

Clearly, this system is integrable for p, and the solution is of the form

$$p = k(x), \tag{25.12}$$

where k is a certain function of x. Under the equilibrating pressure field p given by (25.12) the components of the stress tensor field are of the form

$$\begin{aligned} T_{\langle xx\rangle} &= c, & T_{\langle yy\rangle} &= f(x), & T_{\langle zz\rangle} &= g(x), \\ T_{\langle xy\rangle} &= 0, & T_{\langle xz\rangle} &= 0, & T_{\langle yz\rangle} &= h(x), \end{aligned} \tag{25.13}$$

where c is a constant, and where f, g, h are certain functions of x.

Note. The precise forms of the functions k, f, g, h depend on the component functions $S_{\langle xx\rangle}$, $S_{\langle yy\rangle}$, $S_{\langle zz\rangle}$, and $S_{\langle yz\rangle}$. The important condition here is that the stress tensor field with components given by (25.13) satisfies the equations of equilibrium regardless of what forms the functions f, g, h have. Also, the value of the constant c in (25.13) is entirely arbitrary. This fact reflects the usual condition that an equilibrium configuration of an incompressible material body is not affected by any additional constant pressure field.

25.2. Cylindrical Problems

We consider two types of stress tensor fields. First, we assume that the physical components of the determinate stress $\mathbf{S}$ relative to a cylindrical system (r, θ, z) satisfy the following conditions: (a) the shear components

$S_{\langle r\theta\rangle}$ and $S_{\langle rz\rangle}$ vanish, and (b) the remaining nonzero components of $\mathbf{S}$ depend only on r. Under these conditions the equations of equilibrium reduce to

$$\frac{\partial p}{\partial r} = \frac{dS_{\langle rr\rangle}}{dr} + \frac{1}{r}(S_{\langle rr\rangle} - S_{\langle\theta\theta\rangle}), \qquad \frac{\partial p}{\partial \theta} = 0, \qquad \frac{\partial p}{\partial z} = 0. \tag{25.14}$$

Clearly, this system is integrable for p, and the solution is of the form

$$p = k(r). \tag{25.15}$$

Then the physical components of the stress tensor field are

$$\begin{aligned} T_{\langle rr\rangle} &= f(r), & T_{\langle\theta\theta\rangle} &= \frac{d}{dr}\big(rf(r)\big), & T_{\langle zz\rangle} &= g(r), \\ T_{\langle r\theta\rangle} &= 0, & T_{\langle rz\rangle} &= 0, & T_{\langle\theta z\rangle} &= h(r). \end{aligned} \tag{25.16}$$

As before the functions k, f, g, h depend on the component functions $S_{\langle rr\rangle}$, $S_{\langle\theta\theta\rangle}$, $S_{\langle zz\rangle}$, and $S_{\langle\theta z\rangle}$. However, regardless of what forms f, g, h have, the stress tensor with components given by (25.16) always satisfies the equations of equilibrium.

Next, we consider another type of stress tensor field. We assume that the physical components of the determinate stress $\mathbf{S}$ relative to the cylindrical system satisfy the following conditions: (a) the shear components $S_{\langle rz\rangle}$ and $S_{\langle\theta z\rangle}$ vanish, and (b) the remaining nonzero components of $\mathbf{S}$ are all constant. Under these conditions the equations of equilibrium reduce to

$$\frac{\partial p}{\partial r} = \frac{S_{\langle rr\rangle} - S_{\langle\theta\theta\rangle}}{r}, \qquad \frac{\partial p}{\partial \theta} = 0, \qquad \frac{\partial p}{\partial z} = 0. \tag{25.17}$$

Once again, this system is integrable, and the pressure field is of the form

$$p = (S_{\langle rr\rangle} - S_{\langle\theta\theta\rangle}) \log r + 2\theta S_{\langle r\theta\rangle} + c_0, \tag{25.18}$$

where c_0 is a constant. Under this equilibrating pressure field the stress tensor field has the physical components

$$\begin{aligned} T_{\langle rr\rangle} &= -p + c_1, & T_{\langle\theta\theta\rangle} &= -p + c_2, & T_{\langle zz\rangle} &= -p + c_3, \\ T_{\langle r\theta\rangle} &= c_4, & T_{\langle rz\rangle} &= 0, & T_{\langle\theta z\rangle} &= 0, \end{aligned} \tag{25.19}$$

where c_1, c_2, c_3, c_4 are certain constants. As before the stress tensor field with component fields given by (25.19) always satisfies the equations of equilibrium regardless of what values the constants c_1, c_2, c_3, and c_4 are.

25.3. Spherical Problems

We assume that the physical components of the determinate stress S relative to the spherical system (r, θ, φ) satisfy the following conditions: (a) all shear components $S_{\langle r\theta\rangle}$, $S_{\langle r\varphi\rangle}$, $S_{\langle\theta\varphi\rangle}$ vanish, (b) the normal components $S_{\langle\theta\theta\rangle}$ and $S_{\langle\varphi\varphi\rangle}$ are equal, and (c) all normal components depend on r only. Under these conditions the equations of equilibrium reduce to

$$\frac{\partial p}{\partial r} = \frac{dS_{\langle rr\rangle}}{dr} + \frac{2}{r}(S_{\langle rr\rangle} - S_{\langle\theta\theta\rangle}), \qquad \frac{\partial p}{\partial \theta} = 0, \qquad \frac{\partial p}{\partial \varphi} = 0. \tag{25.20}$$

This system is integrable for p, and the solution is of the form

$$p = k(r). \tag{25.21}$$

The physical components of the stress field are

$$\begin{aligned} T_{\langle rr\rangle} &= f(r), \qquad T_{\langle\theta\theta\rangle} = T_{\langle\varphi\varphi\rangle} = \frac{1}{2r}\frac{d}{dr}(r^2 f(r)), \\ T_{\langle r\theta\rangle} &= 0, \qquad T_{\langle r\varphi\rangle} = 0, \qquad T_{\langle\theta\varphi\rangle} = 0. \end{aligned} \tag{25.22}$$

This field satisfies the equations of equilibrium regardless of what form the function $f(r)$ is.

The preceding static results show that there are several types of stress tensor fields which involve certain functions and constants such that these fields always satisfy the equations of equilibrium regardless of what forms or values the functions or the constants are. Using these results, we can verify easily that the following families of inhomogeneous deformations are universal solutions.

Family 1. The deformation functions are

$$r = (2AX)^{1/2}, \qquad \theta = BY + CZ, \qquad z = DY + EZ, \tag{25.23}$$

where A, B, C, D, E are constants satisfying the condition

$$A(BE - CD) = 1. \tag{25.24}$$

In (25.23) we have used a Cartesian system (X, Y, Z) in the reference configuration $\boldsymbol{\varkappa}$ and a cylindrical system (r, θ, z) in the deformed configuration $\boldsymbol{\chi}$.

We can prove that the deformations in this family are universal solutions by showing that the stress tensor field in the deformed configuration $\boldsymbol{\chi}$ is of the first type in the cylindrical problems. Indeed, from (25.23) the component matrix of the deformation gradient **F** relative to the natural bases of (X, Y, Z) and (r, θ, z) is

$$[F_A{}^i] = \begin{bmatrix} \dfrac{A}{r} & 0 & 0 \\ 0 & B & C \\ 0 & D & E \end{bmatrix}. \tag{25.25}$$

Then the contravariant component matrix of the left Cauchy–Green tensor **B** in the cylindrical system (r, θ, z) is

$$[B^{km}] = [F_A{}^k F_A{}^m] = \begin{bmatrix} \dfrac{A^2}{r^2} & 0 & 0 \\ 0 & B^2 + C^2 & BD + CE \\ 0 & BD + CE & D^2 + E^2 \end{bmatrix}. \tag{25.26}$$

Now using the transformation law (46.19) in Section 46, IVT-2, we obtain the physical component matrix of **B**:

$$[B_{\langle km\rangle}] = [(g_{kk} g_{mm})^{1/2} B^{km}] = \begin{bmatrix} \dfrac{A^2}{r^2} & 0 & 0 \\ 0 & r^2(B^2 + C^2) & r(BD + CE) \\ 0 & r(BD + CE) & D^2 + E^2 \end{bmatrix}, \tag{25.27}$$

where the repeated indices k and m are not summed. From (25.27) we can prove easily that $\det \mathbf{B} = 1$ if and only if (25.24) holds. Thus the deformations in this family are all isochoric.

From (25.27) and (25.24) the physical component matrix of $\mathbf{B}^{-1}$ in (r, θ, z) is

$$[(B^{-1})_{\langle km\rangle}] = \begin{bmatrix} \dfrac{r^2}{A^2} & 0 & 0 \\ 0 & \dfrac{A^2}{r^2}(D^2 + E^2) & -\dfrac{A^2}{r^2}(BD + CE) \\ 0 & -\dfrac{A^2}{r^2}(BD + CE) & A^2(B^2 + C^2) \end{bmatrix}. \tag{25.28}$$

The fundamental invariants I and II can now be read off from (25.27) and (25.28), viz.,

$$\begin{aligned} \mathrm{I} &= \frac{A^2}{r^2} + r^2(B^2 + C^2) + D^2 + E^2, \\ \mathrm{II} &= \frac{r^2}{A^2} + \frac{A^2}{r^2}(D^2 + E^2) + A^2(B^2 + C^2). \end{aligned} \tag{25.29}$$

Substituting (25.27), (25.28), and (25.29) into the representation (25.4), we see that the determinate stress **S** satisfies the conditions (a) and (b) of the first type in the cylindrical problems. Hence the deformed configurations $\boldsymbol{\chi}$ may be maintained in equilibrium by boundary forces alone, and thus the deformations are universal solutions.

Family 2. The deformation functions are

$$x = \tfrac{1}{2}AR^2, \qquad y = B\Theta + CZ, \qquad z = D\Theta + EZ, \tag{25.30}$$

where $A, \ldots, E$ are constants satisfying (25.24). In (25.30) we have used a cylindrical system (R, Θ, Z) in $\boldsymbol{\varkappa}(\mathscr{B})$ and a Cartesian system (x, y, z) in $\boldsymbol{\chi}(\mathscr{B})$.

Following essentially the same procedure as for the preceding family, we can show that the determinate stress **S** in the deformed configurations of this family satisfies the conditions (a) and (b) of the planar problems. Hence the deformations are universal solutions.

Family 3. The deformation functions are

$$r = (AR^2 + B)^{1/2}, \qquad \theta = C\Theta + DZ, \qquad z = E\Theta + FZ. \tag{25.31}$$

For this family we have used the cylindrical system in both the reference configuration $\boldsymbol{\varkappa}(\mathscr{B})$ and the deformed configuration $\boldsymbol{\chi}(\mathscr{B})$.

We can show that the determinate stress **S** in the deformed configuration satisfies the conditions (a) and (b) of the first type in the cylindrical problems. Hence the deformations are universal solutions.

Family 4. The deformation functions are

$$r = (\pm R^3 + A)^{1/3}, \qquad \theta = \pm\Theta, \qquad \varphi = \Phi, \tag{25.32}$$

where A is an arbitrary constant. For this family we have used the spherical coordinate system in both the reference configuration $\boldsymbol{\varkappa}(\mathscr{B})$ and the deformed configuration $\boldsymbol{\chi}(\mathscr{B})$.

We can show that the determinate stress **S** in the deformed configuration satisfies the conditions (a), (b), and (c) of the spherical problems. Hence the deformations are universal solutions.

Family 5. The deformation functions are

$$r = AR, \qquad \theta = B \log R + C\Theta, \qquad z = DZ, \tag{25.33}$$

where $A, \ldots, D$ are constants satisfying the condition

$$A^2CD = 1. \tag{25.34}$$

For this family we have used the cylindrical coordinate system in both $\boldsymbol{\varkappa}(\mathscr{B})$ and $\boldsymbol{\chi}(\mathscr{B})$.

We can show that the determinate stress S in the deformed configuration $\boldsymbol{\chi}(\mathscr{B})$ satisfies the conditions (a) and (b) of the second type in the cylindrical problems. Hence the deformations are universal solutions.

The preceding five families of inhomogeneous universal solutions were discovered originally by Rivlin,[12] Ericksen,[13] Rivlin and Ericksen,[14] and Singh and Pipkin.[15]

Having summarized the presently known families of static universal solutions for incompressible isotropic elastic materials, we consider next the problem of dynamic universal solutions. Using the argument as before [cf. (24.39) and (24.40)], we can show that, in general, a motion is a dynamic universal solution of any type of incompressible elastic materials if and only if it is circulation preserving and its instantaneous deformations are all static universal solutions. In particular, for incompressible isotropic elastic materials, a typical dynamic universal solution may be obtained by replacing the constants $A, B, \ldots$ in the preceding families of static universal solutions by some appropriate functions of time $A(t), B(t), \ldots$ such that the acceleration field of the resulting motion is a conservative field.

Since a complete analysis of such dynamic universal solutions is lengthy, we shall illustrate the results by motions belonging to Family 2 only. For this family the components of the acceleration field are the same as those of a homogeneous motion, since the coordinates (x, y, z) in the

[12] R. S. Rivlin, Torsion of a rubber cylinder, *Journal of Applied Physics*, Vol. 18, pp. 444–449, 1947; Large elastic deformations of isotropic materials, IV. Further developments of the general theory, *Philosophical Transactions of the Royal Society of London A*, Vol. 241, pp. 379–397, 1948; Large elastic deformations of isotropic elastic materials, V. The Problem of flexure, *Proceedings of the Royal Society of London A*, Vol. 195, pp. 463–473, 1949; Large elastic deformations of isotropic elastic materials, VI. Further results in the theory of torsion, shear, and flexure, *Philosophical Transactions of the Royal Society of London A*, Vol. 242, pp. 173–195, 1949.

[13] J. L. Ericksen, Deformation possible in every isotropic, incompressible, perfectly elastic body, *Zeitschrift für Angewandte Mathematik und Physik*, Vol. 5, pp. 466–489, 1954.

[14] R. S. Rivlin and J. L. Ericksen, Stess-deformation relations for isotropic materials, *Journal of Rational Mechanics and Analysis*, Vol. 4, pp. 323–425, 1955. This paper and the preceding ones are reprinted in *Continuum Mechanics*, International Science Review Series, Vol. 8, Edited by C. Truesdell, Gordon and Breach, New York, 1965.

[15] M. Singh and A. C. Pipkin, Note on Ericksen's problem, *Zeitschrift für Angewandte Mathematik und Physik*, Vol. 16, pp. 706–709, 1965.

deformation functions (25.30) are Cartesian. Specifically, we have

$$\begin{aligned} a_x &= \ddot{x} = \frac{\ddot{A}}{A} x, \\ a_y &= \ddot{y} = A(E\ddot{B} - D\ddot{C})y + A(B\ddot{C} - C\ddot{B})z, \\ a_z &= \ddot{z} = A(E\ddot{D} - D\ddot{E})y + A(B\ddot{E} - C\ddot{D})z, \end{aligned} \tag{25.35}$$

where the functions $A(t), \ldots, E(t)$ must satisfy the condition (25.24). We now determine the forms of these functions in order that the acceleration field be conservative; i.e., there is an acceleration potential $\zeta = \zeta(x, y, z, t)$ such that

$$\frac{\partial \zeta}{\partial x} = a_x, \qquad \frac{\partial \zeta}{\partial y} = a_y, \qquad \frac{\partial \zeta}{\partial z} = a_z. \tag{25.36}$$

From (25.35) we see that the condition of integrability for the system (25.36) is

$$B\ddot{C} - C\ddot{B} = E\ddot{D} - D\ddot{E}. \tag{25.37}$$

This differential equation may be integrated with respect to t, yielding

$$B\dot{C} - C\dot{B} = E\dot{D} - D\dot{E} + k, \tag{25.38}$$

where k is a constant. We consider the following two cases:

1. $B(t_0) \neq 0$ at some t_0. In this case the complete solution near t_0 is

$B, D, E =$ arbitrary functions of t, but $B \neq 0$ and E, D do not vanish simultaneously;

$C = B\left[k' + \int_{t_0}^{t} \frac{1}{B^2}(ED' - DE' + k)\,dt\right]$, where k' is a constant such that $E(t_0) - k'D(t_0) = 0$:

$$A = \frac{1}{BE - CD}.$$

2. $B(t_0) = 0$ at some t_0. In this case $C(t_0)$ and $D(t_0)$ do not vanish. The complete solution near t_0 is

$C, D, E =$ arbitrary functions of t but $CD \neq 0$,

$$B = -C \int_{t_0}^{t} \frac{1}{C^2}(E\dot{D} - D\dot{E} + k)\,dt,$$

$$A = \frac{1}{BE - CD}.$$

In general a solution may belong to different cases on different intervals of time. The acceleration potential ζ is given by

$$\zeta = \frac{1}{2}\frac{\ddot{A}}{A}x^2 + \frac{1}{2}A(E\ddot{B} - D\ddot{C})y^2 + \frac{1}{2}A(B\ddot{E} - C\ddot{D})z^2$$
$$+ A(B\ddot{C} - C\ddot{B})yz + f(t), \tag{25.39}$$

where f is an arbitrary function of t.

The preceding dynamic universal solutions were discovered originally by Truesdell.(16)

26. Materially Uniform Smooth Elastic Bodies

In Sections 22–25 we have formulated some mathematical models of certain homogeneous material bodies, and on the basis of these models we have discussed several static and dynamic problems. Now we generalize the models to include certain inhomogeneous material bodies, called *materially uniform smooth bodies*. Originally, the more general models were developed by Noll(17) and by Wang(18) for bodies made up of simple materials. Here, however, we present the models for elastic bodies only, since the major applications of the models are in the theory of elasticity.

We recall that the mechanical response of an elastic material point X in a body manifold $\mathscr{B}$ is characterized by a constitutive equation of the form

$$\mathbf{T} = \mathbf{G}(\mathbf{F}, \mathbf{X}), \tag{26.1}$$

where $\mathbf{X}$ is the position occupied by X in a particular reference configuration $\boldsymbol{\varkappa}$ of $\mathscr{B}$, and where $\mathbf{F}$ is the deformation gradient taken with respect to

(16) C. Truesdell, Solutio Generalis et Accurata Problematum Quamplurimorum de Motu Corporum Elasticorum Incomprimibilium in Deformationibus valde Magnis, *Archive for Rational Mechanics and Analysis*, Vol. 11, pp. 106–113, 1962; Addendum, *Archive for Rational Mechanics and Analysis*, Vol. 12, p. 106, 1963; Corrigenda Addendumque Alterum, *Archive for Rational Mechanics and Analysis*, Vol. 28, p. 397, 1968.

(17) W. Noll, Materially uniform simple bodies with inhomogeneities, *Archive for Rational Mechanics and Analysis*, Vol. 27, pp. 1–32, 1967.

(18) C.-C. Wang, On the geometric structures of simple bodies, a mathematical foundation for the theory of continuous distributions of dislocations, *Archive for Rational Mechanics and Analysis*, Vol. 27, pp. 33–94, 1967. Both this paper and the preceding one are reprinted in W. Noll, R. A. Toupin, and C.-C. Wang, *Continuum Theory of Inhomogeneities in Simple Bodies*, Springer, Berlin, 1968.

$\varkappa(\mathscr{B})$. Suppose that the response function $\mathbf{G}$ is independent of the position $\mathbf{X}$. Then $\mathscr{B}$ is a homogeneous body, and $\varkappa$ is a homogeneous configuration of $\mathscr{B}$. We call $\mathscr{B}$ an *inhomogeneous body* if it does not possess any homogeneous configuration. In this case the response function $\mathbf{G}$ must depend on the position $\mathbf{X}$ relative to any reference configuration $\varkappa$.

Now we consider a special type of inhomogeneous elastic material bodies. For these bodies we assume that the response function $\mathbf{G}$ satisfies the following two basic conditions:

(i) Material Uniformity. There is a *nominal response function* $\mathbf{N}(\mathbf{F})$ such that $\mathbf{G}(\mathbf{F}, \mathbf{X})$ may be represented by

$$\mathbf{G}(\mathbf{F}, \mathbf{X}) = \mathbf{N}(\mathbf{F}\mathbf{K}(\mathbf{X})) \tag{26.2}$$

for all $\mathbf{X} \in \varkappa(\mathscr{B})$ and for all deformation gradients $\mathbf{F}$, where $\mathbf{K}(\mathbf{X})$ is a nonsingular tensor which generally depends on $\mathbf{X}$.

Notice that the nominal response function $\mathbf{N}(\mathbf{F})$ does not depend explicitly on $\mathbf{X}$, so the representation (26.2) requires that the dependence of $\mathbf{G}$ on $\mathbf{X}$ be contained entirely in the term $\mathbf{K}(\mathbf{X})$. Furthermore, if we write the representation (26.2) as

$$\mathbf{G}(\mathbf{F}\mathbf{K}^{-1}(\mathbf{X}), \mathbf{X}) = \mathbf{N}(\mathbf{F}), \tag{26.3}$$

we see that $\mathbf{K}^{-1}(\mathbf{X})$ is a tensor which transforms the response function at $\mathbf{X}$ to the nominal response function $\mathbf{N}(\mathbf{F})$ for all $\mathbf{X} \in \varkappa(\mathscr{B})$. In other words, by applying $\mathbf{K}^{-1}(\mathbf{X})$ to the tangent space at $\mathbf{X}$, we can map the state of X in $\varkappa(\mathscr{B})$ to a state with response function $\mathbf{N}(\mathbf{F})$. Since $\mathbf{N}(\mathbf{F})$ is independent of X, (26.3) means also that $\mathbf{K}^{-1}(\mathbf{X})$ and $\mathbf{K}^{-1}(\mathbf{Y})$ map the states of X and Y in $\varkappa(\mathscr{B})$ to the same state; i.e.,

$$\mathbf{G}(\mathbf{F}\mathbf{K}^{-1}(\mathbf{X}), \mathbf{X}) = \mathbf{G}(\mathbf{F}\mathbf{K}^{-1}(\mathbf{Y}), \mathbf{Y}) \tag{26.4}$$

for all deformation gradients $\mathbf{F}$.

In general if the condition (26.4) holds for some tensors $\mathbf{K}^{-1}(\mathbf{X})$ and $\mathbf{K}^{-1}(\mathbf{Y})$, then we say that X and Y are *materially isomorphic*. Since (26.3) is equivalent to (26.4) for all pairs X, Y in $\mathscr{B}$, the condition of material uniformity corresponds precisely to the requirement that the body points of $\mathscr{B}$ are pairwise materially isomorphic.

Now it should be noted that, in general, the tensor $\mathbf{K}(\mathbf{X})$ in (26.2) is not unique. To see this fact, we define the *nominal symmetry group* $\mathscr{G}$ by

$$\mathbf{H} \in \mathscr{G} \Leftrightarrow \mathbf{N}(\mathbf{F}\mathbf{H}) = \mathbf{N}(\mathbf{F}) \qquad \text{for all } \mathbf{F}. \tag{26.5}$$

Then we have

$$\mathbf{N}(\mathbf{F}\mathbf{K}(\mathbf{X})) = \mathbf{N}(\mathbf{F}\mathbf{K}(\mathbf{X})\mathbf{H}(\mathbf{X})) \tag{26.6}$$

for any $\mathbf{H}(\mathbf{X}) \in \mathscr{G}$. As a result, if $\mathbf{K}(\mathbf{X})$ satisfies (26.2), then $\mathbf{K}(\mathbf{X})\mathbf{H}(\mathbf{X})$ also satisfies (26.2) for all $\mathbf{H}(\mathbf{X}) \in \mathscr{G}$. Hence as long as $\mathscr{G}$ is not the trivial group formed by $\mathbf{I}$ alone, the tensor $\mathbf{K}(\mathbf{X})$ which satisfies (26.2) is not unique.

Next, we remark that the field $\mathbf{K}(\cdot)$ in (26.2) need not be smooth. Of course, if $\mathbf{K}(\cdot)$ is a smooth field, then $\mathbf{G}$ depends smoothly on $\mathbf{X}$. However, the converse of this assertion is not true. Indeed, since smoothness is a local property, we may characterize the smooth dependence of $\mathbf{G}$ on $\mathbf{X}$ by the following condition:

(ii) Smoothness. For each $\mathbf{X}_0 \in \boldsymbol{\varkappa}(\mathscr{B})$ we can choose a neighborhood $\mathscr{N}_0$ and a smooth nonsingular tensor field $\mathbf{K}$ on $\mathscr{N}_0$ such that the representation (26.2) holds for all $\mathbf{X}$ in $\mathscr{N}_0$.

If $\bar{\mathbf{X}}_0$ is not contained in the neighborhood $\mathscr{N}_0$ of $\mathbf{X}_0$, then by the same smoothness condition we can choose a neighborhood $\bar{\mathscr{N}}_0$ of $\bar{\mathbf{X}}_0$ and a field $\bar{\mathbf{K}}$ on $\bar{\mathscr{N}}_0$ such that the representation

$$\mathbf{G}(\mathbf{F}, \mathbf{X}) = \mathbf{N}(\mathbf{F}\bar{\mathbf{K}}(\mathbf{X})) \tag{26.7}$$

holds for all $\mathbf{X} \in \bar{\mathscr{N}}_0$. In particular, if $\mathscr{N}_0$ overlaps $\bar{\mathscr{N}}_0$, and if $\mathbf{X} \in \mathscr{N}_0 \cap \bar{\mathscr{N}}_0$, then both (26.2) and (26.7) are satisfied. Hence

$$\mathbf{N}(\mathbf{F}\mathbf{K}(\mathbf{X})) = \mathbf{N}(\mathbf{F}\bar{\mathbf{K}}(\mathbf{X})) \tag{26.8}$$

for all deformation gradients $\mathbf{F}$. Comparing this condition with (26.5), we see that the difference

$$\mathbf{H}(\mathbf{X}) = \mathbf{K}^{-1}(\mathbf{X})\bar{\mathbf{K}}(\mathbf{X}) \tag{26.9}$$

between $\mathbf{K}(\mathbf{X})$ and $\bar{\mathbf{K}}(\mathbf{X})$ is contained in $\mathscr{G}$. Thus the smooth fields $\mathbf{K}$ and $\bar{\mathbf{K}}$ are related to each other by a smooth field $\mathbf{H}$ on $\mathscr{N}_0 \cap \bar{\mathscr{N}}_0$ with values belonging to the nominal symmetry group $\mathscr{G}$.

If $\mathbf{H}(\mathbf{X}) = \mathbf{I}$ for all $\mathbf{X} \in \mathscr{N}_0 \cap \bar{\mathscr{N}}_0$, then $\mathbf{K}$ and $\bar{\mathbf{K}}$ coincide on the overlap of their domains. In this case we can regard $\bar{\mathbf{K}}$ as an extension of $\mathbf{K}$, and vice versa. In general $\mathbf{H}(\mathbf{X})$ need not coincide with $\mathbf{I}$, of course. Then $\mathbf{K}$ and $\bar{\mathbf{K}}$ are not the same field on $\mathscr{N}_0 \cap \bar{\mathscr{N}}_0$, but they both satisfy the representation (26.2) or (26.7).

We remark that a smooth field $\mathbf{K}^{-1}$ satisfying (26.3) need not be the gradient of a deformation. Indeed, if $\mathbf{K}^{-1} = \operatorname{grad} \boldsymbol{\varphi}$, then the configuration

$\bar{\varkappa} = \boldsymbol{\varphi} \circ \boldsymbol{\varkappa}$ is homogeneous for $\mathscr{B}$. This assertion is obvious, since by the transformation law (18.24) of the response function the left-hand side of (26.3) is just the response function $\bar{\mathbf{G}}$ relative to the configuration $\bar{\varkappa}$; i.e.,

$$\bar{\mathbf{G}}(\mathbf{F}, \boldsymbol{\varphi}(\mathbf{X})) = \mathbf{G}(\mathbf{F} \operatorname{grad} \boldsymbol{\varphi}, \mathbf{X}). \tag{26.10}$$

Then (26.3) implies that $\bar{\mathbf{G}}$ coincides with $\mathbf{N}$. Thus $\bar{\mathbf{G}}$ is independent of the position $\bar{\mathbf{X}} = \boldsymbol{\varphi}(\mathbf{X}) \in \bar{\varkappa}(\mathscr{B})$.

We call $\mathscr{B}$ a *locally homogeneous body* if we can choose the smooth field $\mathbf{K}^{-1}$ on the neighborhood $\mathscr{N}_0$ of $\mathbf{X}_0$ to be the gradient of a deformation of $\mathscr{N}_0$ for each $\mathbf{X}_0 \in \boldsymbol{\varkappa}(\mathscr{B})$. Since $\mathbf{K}$ need not have a smooth extension which is defined on the whole configuration $\boldsymbol{\varkappa}(\mathscr{B})$, and which satisfies the representation (26.2), a locally homogeneous body need not be (globally) homogeneous.

An example of a locally homogeneous body that is not (globally) homogeneous is given by the *Möbius crystal*, which is formed by twisting a (thick) strip of homogeneous crystal into a Möbius strip in the usual way. The resulting Möbius crystal is locally homogeneous, since each segment of the strip may be deformed into a configuration in which the crystalline basis forms a parallel field. Such a configuration is homogeneous for the crystal. The whole Möbius crystal, however, is not homogeneous, since there is no global configuration in which the crystalline basis is a parallel field. In fact for that crystal a globally smooth field $\mathbf{K}$, which satisfies the representation (26.2), does not exist.

In the remainder of this chapter we shall assume that $\mathscr{B}$ is a materially uniform smooth elastic body; i.e., the response function $\mathbf{G}$ of $\mathscr{B}$ satisfies the preceding conditions (i) and (ii). Our main purpose is to develop a theory in order to determine the static and the dynamic response of the body $\mathscr{B}$. A major difficulty in the theory is due to the fact that in general the smooth field $\mathbf{K}$, which satisfies the representation (26.2), is defined only on a neighborhood. To overcome this difficulty, we introduce a concept called a *material connection*. We shall explain this concept in detail in the following section.

27. Material Connections

In Section 26 we explained that the response function at any point $\mathbf{X} \in \boldsymbol{\varkappa}(\mathscr{B})$ may be represented by a nominal response function in the following way:

$$\mathbf{G}(\mathbf{F}, \mathbf{X}) = \mathbf{N}(\mathbf{F}\mathbf{K}(\mathbf{X})), \tag{27.1}$$

where $\mathbf{K}(\mathbf{X})$ is a certain nonsingular tensor depending on $\mathbf{X}$. Since (27.1) holds for all $\mathbf{F}$, it is equivalent to

$$\mathbf{G}(\mathbf{F}\mathbf{K}^{-1}(\mathbf{X}), \mathbf{X}) = \mathbf{N}(\mathbf{F}), \tag{27.2}$$

where the right-hand side no longer depends on $\mathbf{X}$. As a result, if $\mathbf{Y}$ is another point in $\boldsymbol{\varkappa}(\mathscr{B})$, then

$$\mathbf{G}(\mathbf{F}\mathbf{K}^{-1}(\mathbf{Y}), \mathbf{Y}) = \mathbf{G}(\mathbf{F}\mathbf{K}^{-1}(\mathbf{X}), \mathbf{X}). \tag{27.3}$$

Now since (27.3) also holds for all $\mathbf{F}$, it is equivalent to

$$\mathbf{G}(\mathbf{F}, \mathbf{Y}) = \mathbf{G}(\mathbf{F}\mathbf{K}(\mathbf{Y})\mathbf{K}^{-1}(\mathbf{X}), \mathbf{X}), \tag{27.4}$$

which means that the response function at $\mathbf{Y}$ differs from that at $\mathbf{X}$ by a linear transformation

$$\mathbf{K}(\mathbf{X}, \mathbf{Y}) \equiv \mathbf{K}(\mathbf{Y})\mathbf{K}^{-1}(\mathbf{X}) \tag{27.5}$$

from the tangent space at $\mathbf{X}$ to the tangent space at $\mathbf{Y}$.

More specifically, let $\mathscr{V}(\mathbf{X})$ denote the tangent space at any point $\mathbf{X} \in \boldsymbol{\varkappa}(\mathscr{B})$. Then $\mathbf{K}(\mathbf{X}, \mathbf{Y})$ is an isomorphism from $\mathscr{V}(\mathbf{X})$ to $\mathscr{V}(\mathbf{Y})$ such that the stress tensor due to a deformation gradient $\mathbf{F}$ at $\mathbf{Y}$ is always the same as that due to a deformation gradient $\mathbf{F}\mathbf{K}(\mathbf{X}, \mathbf{Y})$ at $\mathbf{X}$ for all choice of $\mathbf{F}$. Because it has this transformation property, $\mathbf{K}(\mathbf{X}, \mathbf{Y})$ is called a *material isomorphism* from $\mathbf{X}$ to $\mathbf{Y}$ relative to $\boldsymbol{\varkappa}(\mathscr{B})$.

In general a material isomorphism from $\mathbf{X}$ to $\mathbf{Y}$ is not unique. Since the composition of any two material isomorphisms is a material isomorphism, when $\mathbf{K}(\mathbf{X}, \mathbf{Y})$ is a material isomorphism from $\mathbf{X}$ to $\mathbf{Y}$, an isomorphism $\bar{\mathbf{K}}(\mathbf{X}, \mathbf{Y})$ is also a material isomorphism if and only if the composition $\mathbf{K}^{-1}(\mathbf{X}, \mathbf{Y})\bar{\mathbf{K}}(\mathbf{X}, \mathbf{Y})$ is contained in the symmetry group of $\mathbf{X}$ [or, equivalently, the composition $\bar{\mathbf{K}}(\mathbf{X}, \mathbf{Y})\mathbf{K}^{-1}(\mathbf{X}, \mathbf{Y})$ is contained in the symmetry group of $\mathbf{Y}$]. Another way to visualize this result is to use the formula (27.5) directly. Since $\mathbf{K}(\mathbf{X})$ and $\mathbf{K}(\mathbf{Y})$ are unique to within a multiple on the right by an arbitrary element in $\mathscr{G}$, a material isomorphism from $\mathbf{X}$ to $\mathbf{Y}$ in general can be represented by

$$\bar{\mathbf{K}}(\mathbf{X}, \mathbf{Y}) = \mathbf{K}(\mathbf{Y})\mathbf{H}\mathbf{K}^{-1}(\mathbf{X}), \tag{27.6}$$

where $\mathbf{H} \in \mathscr{G}$.

It should be noted that the condition of material uniformity is equivalent to the requirement that there be a material isomorphism $\mathbf{K}(\mathbf{X}, \mathbf{Y})$ for each pair of points $\mathbf{X}, \mathbf{Y}$ in $\boldsymbol{\varkappa}(\mathscr{B})$. Indeed, choosing any reference point

$\mathbf{X}_0 \in \varkappa(\mathscr{B})$, we can regard $\mathbf{G}(\mathbf{F}, \mathbf{X}_0)$ as a nominal response function. Then the transformation rule

$$\mathbf{G}(\mathbf{F}, \mathbf{X}) = \mathbf{G}\big(\mathbf{F}\mathbf{K}(\mathbf{X}_0, \mathbf{X}), \mathbf{X}_0\big) \tag{27.7}$$

becomes a representation of $\mathbf{G}(\mathbf{F}, \mathbf{X})$ by means of the nominal response function $\mathbf{N}(\mathbf{F}) \equiv \mathbf{G}(\mathbf{F}, \mathbf{X}_0)$ and the tensor field $\mathbf{K}(\mathbf{X}) \equiv \mathbf{K}(\mathbf{X}_0, \mathbf{X})$.

By the same token the condition of smoothness is equivalent to the requirement that for each point $\mathbf{X}_0 \in \varkappa(\mathscr{B})$ we can choose a neighborhood $\mathscr{N}_0$ and a smooth field of material isomorphisms $\mathbf{K}(\mathbf{X}_0, \mathbf{X})$ for all $\mathbf{X} \in \mathscr{N}_0$. We can regard such a smooth field as a parallelism from the point $\mathbf{X}_0$ to neighboring points $\mathbf{X} \in \mathscr{N}_0$. A globally smooth field of material isomorphisms $\mathbf{K}(\mathbf{X}_0, \mathbf{X})$ for all $\mathbf{X} \in \varkappa(\mathscr{B})$ may or may not exist, however. Hence the geometric structure on $\varkappa(\mathscr{B})$ defined by the material isomorphisms need not give rise to a distant parallelism.

In Sections 56 and 64, IVT-2, we have discussed two types of parallelisms on a manifold, namely, the Riemannian parallelism induced by a metric and the Cartan parallelism defined on a continuous group. We have remarked that the former may have curvature but not torsion, while the latter may have torsion but not curvature. For the purpose of characterizing material isomorphisms, we need a more general type of parallelism which may have both curvature and torsion. In classical differential geometry this general type of parallelism is called an *affine parallelism* or an *affine connection*, which we shall now explain.

Let $\{\mathbf{h}_A\}$ be the natural basis field of the coordinate system (X^A) on $\varkappa(\mathscr{B})$ as before. We define a general covariant derivative of $\{\mathbf{h}_A\}$ by

$$\frac{D\mathbf{h}_A}{DX^B} = \Gamma^C_{AB}\,\mathbf{h}_C. \tag{27.8}$$

This formula is formally the same as the formula (56.5) in Section 56, IVT-2, except that the Christoffel symbols are replaced by the Γ symbols Γ^C_{AB}. In general, Γ^C_{AB} need not be obtained from a Riemannian metric. We simply assign the Γ symbols as certain smooth functions of (X^D). In particular, Γ^C_{AB} need not be symmetric with respect to the subscripts A and B.

Using the basic formula (27.8), we can define the covariant derivative of a vector field $\mathbf{u} = u^A\mathbf{h}_A$ in general by the component formula

$$\nabla\mathbf{u} = \left(\frac{\partial u^C}{\partial X^B} + u^A\Gamma^C_{AB}\right)\mathbf{h}_C \otimes \mathbf{h}^B. \tag{27.9}$$

In particular, Γ^C_{AB} are also the components of $\nabla \mathbf{h}_A$, viz.,

$$\nabla \mathbf{h}_A = \Gamma^C_{AB} \mathbf{h}_C \otimes \mathbf{h}^B. \tag{27.10}$$

Now consider a smooth curve $\boldsymbol{\lambda}(\tau)$ with coordinates $(\lambda^A(\tau))$ relative to the coordinate system (X^A). Suppose that $\mathbf{u} = \mathbf{u}(\tau)$ is a vector field on $\boldsymbol{\lambda}(\tau)$. Then we define the covariant derivative of $\mathbf{u}$ along the curve $\boldsymbol{\lambda}$ by

$$\frac{D\mathbf{u}}{D\tau} \equiv \left(\frac{du^C}{d\tau} + u^A \Gamma^C_{AB} \frac{d\lambda^B}{d\tau} \right) \mathbf{h}_C. \tag{27.11}$$

As before we call $\mathbf{u}(\tau)$ a *parallel field* or a *covariantly constant field* along $\boldsymbol{\lambda}(\tau)$ if $D\mathbf{u}/D\tau = \mathbf{0}$. Thus in component form the equations of parallel transport are

$$\frac{du^C}{d\tau} + u^A \Gamma^C_{AB} \frac{d\lambda^B}{d\tau} = 0, \qquad C = 1, 2, 3. \tag{27.12}$$

The parallelism defined by these equations is the affine parallelism induced by the Γ symbols Γ^C_{AB} in the (X^A) coordinate system.

Notice that the formulas (27.8)–(27.12) associated by the Γ symbols are formally the same as those in Sections 56 and 64, IVT-2, for the Christoffel symbols and the Cartan symbols. Hence an affine parallelism is a direct generalization of a Riemannian parallelism or a Cartan parallelism. As usual we define the torsion tensor $\boldsymbol{\Upsilon}$ by the component formula

$$\boldsymbol{\Upsilon} \equiv (\Gamma^C_{AB} - \Gamma^C_{BA}) \mathbf{h}_C \otimes \mathbf{h}^A \otimes \mathbf{h}^B, \tag{27.13}$$

and the curvature tensor $\mathbf{R}$ by the component formula

$$\mathbf{R} = \left(\frac{\partial \Gamma^C_{AB}}{\partial X^D} - \frac{\partial \Gamma^C_{AD}}{\partial X^B} + \Gamma^E_{AB} \Gamma^C_{ED} - \Gamma^E_{AD} \Gamma^C_{EB} \right) \mathbf{h}_C \otimes \mathbf{h}^A \otimes \mathbf{h}^D \otimes \mathbf{h}^B. \tag{27.14}$$

Then in general neither $\boldsymbol{\Upsilon}$ nor $\mathbf{R}$ vanishes.

So far, we have defined the affine parallelism and its related quantities and operations in the (global) coordinate system (X^A). Now suppose that $(\bar{X}^A)$ is another coordinate system on $\boldsymbol{\varkappa}(\mathscr{B})$, and let the coordinate transformation from (X^A) to $(\bar{X}^A)$ be given by $X^A = X^A(\bar{X}^B)$ and $\bar{X}^A = \bar{X}^A(X^B)$. Then as explained in Section 45, IVT-2, the natural basis field $\{\bar{\mathbf{h}}_A\}$ of $(\bar{X}^A)$ is related to the natural basis field $\{\mathbf{h}_A\}$ of (X^A) by

$$\bar{\mathbf{h}}_A = \frac{\partial X^B}{\partial \bar{X}^A} \mathbf{h}_B. \tag{27.15}$$

Hence the covariant derivative of $\bar{\mathbf{h}}_A$ relative to the affine parallelism under consideration is given by the formula (27.9), viz.,

$$\nabla \bar{\mathbf{h}}_A = \left(\frac{\partial^2 X^C}{\partial \bar{X}^A \, \partial \bar{X}^D} \frac{\partial \bar{X}^D}{\partial X^B} + \frac{\partial X^D}{\partial \bar{X}^A} \Gamma^C_{DB} \right) \mathbf{h}_C \otimes \mathbf{h}^B \qquad (27.16)$$

or, equivalently, in component form relative to $\{\bar{\mathbf{h}}_C\}$

$$\nabla \bar{\mathbf{h}}_A = \left(\frac{\partial^2 X^C}{\partial \bar{X}^A \, \partial \bar{X}^D} \frac{\partial \bar{X}^D}{\partial X^B} + \frac{\partial X^D}{\partial \bar{X}^A} \Gamma^C_{DB} \right) \frac{\partial \bar{X}^E}{\partial X^C} \frac{\partial X^B}{\partial \bar{X}^F} \bar{\mathbf{h}}_E \otimes \bar{\mathbf{h}}^F. \qquad (27.17)$$

Comparing this formula with (27.10), we see that the Γ symbols in the $(\bar{X}^D)$ coordinate system are given by

$$\bar{\Gamma}^E_{AF} = \Gamma^C_{DB} \frac{\partial X^D}{\partial \bar{X}^A} \frac{\partial \bar{X}^E}{\partial X^C} \frac{\partial X^B}{\partial \bar{X}^F} + \frac{\partial^2 X^C}{\partial \bar{X}^A \, \partial \bar{X}^F} \frac{\partial \bar{X}^E}{\partial X^C}. \qquad (27.18)$$

This formula characterizes the transformation law for the Γ symbols of a particular affine connection under a change of coordinate system. Using (27.18), we can prove by direct calculation that the component formulas for the torsion tensor $\mathbf{\Upsilon}$ and the curvature tensor $\mathbf{R}$ are still of the form (27.13) and (27.14) provided that the Γ symbols Γ^C_{AB} and the bases $\{\mathbf{h}_A\}$, $\{\mathbf{h}^A\}$ are replaced by $\bar{\Gamma}^C_{AB}$ and $\{\bar{\mathbf{h}}_A\}$, $\{\bar{\mathbf{h}}^A\}$, respectively.

It should be noted that, under the change of coordinates from (X^A) to $(\bar{X}^A)$ the covariant derivative and the parallel transport both remain unchanged. In other words the tensor field $\nabla \mathbf{u}$ given by the component formula

$$\nabla \mathbf{u} = \left(\frac{\partial \bar{u}^C}{\partial \bar{X}^B} + \bar{u}^A \bar{\Gamma}^C_{AB} \right) \bar{\mathbf{h}}_C \otimes \bar{\mathbf{h}}^B \qquad (27.19)$$

is the same as that given by (27.9). We can prove this fact by the transformation law

$$\left(\frac{\partial u^C}{\partial X^B} + u^A \Gamma^C_{AB} \right) = \left(\frac{\partial \bar{u}^D}{\partial \bar{X}^E} + \bar{u}^F \bar{\Gamma}^D_{FE} \right) \frac{\partial X^C}{\partial \bar{X}^D} \frac{\partial \bar{X}^E}{\partial X^B}, \qquad (27.20)$$

which follows from (27.18) and the usual transformation law

$$u^C = \bar{u}^D \frac{\partial X^C}{\partial \bar{X}^D} \qquad (27.21)$$

for the components of a vector field.

Similarly if $\mathbf{u}(\tau)$ is a parallel field on $\boldsymbol{\lambda}(\tau)$, then the components $\bar{u}^A(\tau)$ relative to $(\bar{X}^A)$ satisfy the equations of parallel transport

$$\frac{d\bar{u}^C}{d\tau} + \bar{u}^A \bar{\Gamma}^C_{AB} \frac{d\bar{\lambda}^B}{d\tau} = 0, \qquad C = 1, 2, 3, \tag{27.22}$$

which are formally the same as (17.12) except that all quantities are taken with respect to the coordinate system $(\bar{X}^A)$. The parallel transport

$$\boldsymbol{\rho}_\tau : \mathscr{V}(\boldsymbol{\lambda}(0)) \to \mathscr{V}(\boldsymbol{\lambda}(\tau)) \tag{27.23}$$

defined by

$$\boldsymbol{\rho}_\tau(\mathbf{u}(0)) \equiv \mathbf{u}(\tau) \tag{27.24}$$

for all parallel fields $\mathbf{u}(\tau)$ on $\boldsymbol{\lambda}(\tau)$, however, remains the same whether $\mathbf{u}(\tau)$ is obtained from (27.12) or from (27.22).

Now we define a material connection on $\boldsymbol{\varkappa}(\mathscr{B})$ as follows: an affine connection such that the parallel transports along all curves are material isomorphisms. It is known that[19] such an affine connection exists. To meet the requirement that the parallel transports be material isomorphisms, the Γ symbols of a material connection are not arbitrary but must satisfy certain conditions, which we proceed to derive.

Consider a neighborhood $\mathscr{N}$ of $\mathbf{X}$ and a smooth field $\mathbf{K}$ on $\mathscr{N}$ satisfying the representation (26.2). Suppose that $\boldsymbol{\lambda}$ is a smooth curve in $\mathscr{N}$, and let $\boldsymbol{\rho}_\tau$ be the parallel transport along $\boldsymbol{\lambda}$ induced by an affine connection. Then as remarked before, $\boldsymbol{\rho}_\tau$ is a material isomorphism if and only if it can be represented by

$$\boldsymbol{\rho}_\tau = \mathbf{K}(\boldsymbol{\lambda}(\tau))\mathbf{H}(\tau)\mathbf{K}^{-1}(\boldsymbol{\lambda}(0)), \tag{27.25}$$

where $\mathbf{H}(\tau) \in \mathscr{G}$. At $\tau = 0$ we have the initial condition

$$\mathbf{H}(0) = \mathbf{I}, \tag{27.26}$$

since $\boldsymbol{\rho}_0$ is the identity map. The representation (27.25) means that a parallel field $\mathbf{u}(\tau)$ on $\boldsymbol{\lambda}(\tau)$ must be of the form

$$\mathbf{u}(\tau) = \mathbf{K}(\boldsymbol{\lambda}(\tau))\mathbf{H}(\tau)\mathbf{K}^{-1}(\boldsymbol{\lambda}(0))\mathbf{u}(0) = \mathbf{K}(\boldsymbol{\lambda}(\tau))\mathbf{H}(\tau)\mathbf{w}, \tag{27.27}$$

where $\mathbf{u}(0)$ and $\mathbf{w}$ are arbitrary vectors in $\mathscr{V}(\boldsymbol{\lambda}(0))$.

(19) C.-C. Wang, On the geometric structures of simple bodies, a mathematical foundation for the theory of continuous distributions of dislocations, *Archive for Rational Mechanics and Analysis*, Vol. 27, pp. 33–94, 1967.

The component form of the vector field $\mathbf{u}(\tau)$ defined by (27.27) is

$$u^C(\tau) = K_B{}^C(\boldsymbol{\lambda}(\tau))H_A{}^B(\tau)w^A. \tag{27.28}$$

Substituting this form into (27.12), we get

$$\frac{dH_A{}^B}{d\tau} K_B{}^C w^A = -\left(\frac{\partial K_B{}^C}{\partial X^D} H_A{}^B + K_B{}^E H_A{}^B \Gamma^C_{ED}\right) \frac{d\lambda^D}{d\tau} w^A. \tag{27.29}$$

Since the vector $\mathbf{w}$ is arbitrary, and since $\mathbf{K}$ is invertible, the preceding condition is equivalent to

$$\frac{dH_A{}^B}{d\tau} = -(K^{-1})_C{}^B\left(\frac{\partial K_F{}^C}{\partial X^D} H_A{}^F + K_F{}^E H_A{}^F \Gamma^C_{ED}\right) \frac{d\lambda^D}{d\tau}. \tag{27.30}$$

In particular, at $\tau = 0$

$$\left.\frac{dH_A{}^B}{d\tau}\right|_0 = -(K^{-1})_C{}^B\left.\left(\frac{\partial K_A{}^C}{\partial X^D} + K_A{}^E \Gamma^C_{ED}\right)\right|_{\boldsymbol{\lambda}(0)} \left.\frac{d\lambda^D}{d\tau}\right|_0. \tag{27.31}$$

It suffices to consider the condition (27.30) at the initial point $\boldsymbol{\lambda}(0)$ only, since that point and the curve $\boldsymbol{\lambda}(\tau)$ passing through it are both arbitrary.

The formula (27.31) shows that the combination

$$(K^{-1})_C{}^B\left.\left(\frac{\partial K_A{}^C}{\partial X^D} + K_A{}^E \Gamma^C_{ED}\right)\right|_{\boldsymbol{\lambda}(0)} \left.\frac{d\lambda^D}{d\tau}\right|_0 \tag{27.32}$$

must be a tangent vector of a curve $\mathbf{H}(\tau) \in \mathscr{G}$ at the identity element $\mathbf{H}(0) = \mathbf{I}$. In Section 65, IVT-2, we have explained that such a tangent vector corresponds to an element in the Lie algebra $\mathscr{g}$ of $\mathscr{G}$. Hence the Γ symbols of a material connection must satisfy the condition that the combination (27.32) is contained in $\mathscr{g}$ for all $\boldsymbol{\lambda}(0) \in \mathscr{N}$ and for all initial tangent vectors $\dot{\boldsymbol{\lambda}}(0) \in \mathscr{V}(\boldsymbol{\lambda}(0))$. As a result, we obtain the following important field condition[20] for the Γ symbols of a material connection:

$$(K^{-1})_C{}^B\left(\frac{\partial K_A{}^C}{\partial X^D} + K_A{}^E \Gamma^C_{ED}\right) \in \mathscr{g}, \tag{27.33}$$

where D is a free index, $D = 1, 2, 3$.

[20] C.-C. Wang, On the geometric structures of simple bodies, a mathematical foundation for the theory of continuous distributions of dislocations, *Archive for Rational Mechanics and Analysis*, Vol. 27, pp. 33–94, 1967.

It should be noted that the condition (27.33) is valid for any smooth field **K** which satisfies the representation (26.2). In application we choose a covering of $\boldsymbol{\varkappa}(\mathscr{B})$ by neighborhoods $\mathscr{N}$ on which we define the smooth fields **K**. Then (27.33) may be used at all points $\mathbf{X} \in \boldsymbol{\varkappa}(\mathscr{B})$ with **K** defined on a neighborhood of **X**.

28. Noll's Equations of Motion

For a homogeneous elastic body the governing equations of the deformation functions may be derived in the following way: We choose a homogeneous reference configuration $\boldsymbol{\varkappa}(\mathscr{B})$, and let **G** be the response function relative to $\boldsymbol{\varkappa}$. Then the stress tensor in the deformed configuration $\boldsymbol{\chi}_t$ is given by the component formula

$$T^{ij} = T^{ij}(\mathbf{x}, t) = G^{ij}\left(\frac{\partial x^k}{\partial X^A}\right). \tag{28.1}$$

Since the stress tensor field must satisfy the linear momentum equation (17.12), we get

$$G^{ij}{}_k{}^A\left(\frac{\partial x^l}{\partial X^C}\right)\frac{\partial^2 x^k}{\partial X^A\,\partial X^B}\frac{\partial X^B}{\partial x^j} + \varrho b^i = \varrho\frac{\partial^2 x^i}{\partial t^2}, \tag{28.2}$$

where $G^{ij}{}_k{}^A(\mathbf{F})$ denotes the gradient of the function $G^{ij}(\mathbf{F})$:

$$G^{ij}{}_k{}^A(\mathbf{F}) = \frac{\partial G^{ij}(\mathbf{F})}{\partial F^k{}_A}. \tag{28.3}$$

The matrix $[\partial X^B/\partial x^j]$ is just the inverse of the deformation gradient $[\partial x^j/\partial X^B]$; it may be absorbed into the coefficient functions $G^{ij}{}_k{}^A(\partial x^l/\partial X^C)$ or, equivalently, (28.3) may be rewritten as

$$H^{iB}{}_k{}^A\left(\frac{\partial x^l}{\partial X^C}\right)\frac{\partial^2 x^k}{\partial X^A\,\partial X^B} + \varrho_\varkappa b^i = \varrho_\varkappa\frac{\partial^2 x^i}{\partial t^2}, \tag{28.4}$$

where $\varrho_\varkappa$ denotes the density in $\boldsymbol{\varkappa}(\mathscr{B})$, and where

$$H^{iB}{}_k{}^A(\mathbf{F}) \equiv (\det\mathbf{F})G^{ij}{}_k{}^A(\mathbf{F})(F^{-1})_j{}^B. \tag{28.5}$$

The system (28.4) governs the deformation functions $x^i(X^A, t)$, $i = 1, 2, 3$. On the basis of the system of equations (28.4) we may formulate various boundary-value or boundary-initial-value problems for the determination of the functions $x^i(X^A, t)$.

A similar system of equations of motion may be derived for a materially uniform smooth elastic body. The concept of a material connection introduced in Section 27 is important in the derivation. Our starting point is the representation (28.2), where $\mathbf{K}$ is a smooth field defined on a neighborhood $\mathscr{N} \subset \boldsymbol{\varkappa}(\mathscr{B})$. In component form the stress tensor is then given by

$$T^{ij}(\mathbf{x}, t) = N^{ij}\left(\frac{\partial x^k}{\partial X^C} K_A{}^C\right). \tag{28.6}$$

Substituting this representation into the linear momentum equation (17.12), we get

$$N^{ij}{}_k{}^A\left(\frac{\partial x^l}{\partial X^E} K_F{}^E\right)\left(K_A{}^C \frac{\partial^2 x^k}{\partial X^C \partial X^D} + \frac{\partial x^k}{\partial X^C} \frac{\partial K_A{}^C}{\partial X^D}\right) \frac{\partial X^D}{\partial x^j} + \varrho b^i = \varrho \frac{\partial^2 x^i}{\partial t^2}, \tag{28.7}$$

where

$$N^{ij}{}_k{}^A(\mathbf{F}) \equiv \frac{\partial N^{ij}(\mathbf{F})}{\partial F_A{}^k}. \tag{28.8}$$

Unlike the system (28.4) for the homogeneous body, the system (28.7) is valid only on the neighborhood $\mathscr{N}$, on which the smooth field $\mathbf{K}$ is defined.

In order that the equations of motion may be written in a global form, we make use of the symmetry condition of the nominal response function $\mathbf{N}$. From (26.5)

$$N^{ij}\left(\frac{\partial x^k}{\partial X^C} K_B{}^C H_A{}^B(\tau)\right) = N^{ij}\left(\frac{\partial x^k}{\partial X^C} K_A{}^C\right) \tag{28.9}$$

for any curve $[H_A{}^B] \in \mathscr{G}$ such that

$$H_A{}^B(0) = \delta_A{}^B. \tag{28.10}$$

Differentiating (28.9) with respect to τ and evaluating the result at $\tau = 0$, we obtain

$$N^{ij}{}_k{}^A\left(\frac{\partial x^l}{\partial X^E} K_F{}^E\right) \frac{\partial x^k}{\partial X^C} K_B{}^C \left.\frac{dH_A{}^B}{d\tau}\right|_0 = 0, \tag{28.11}$$

where the tangent vector of $[H_A{}^B(\tau)]$ at $\tau = 0$ may be an arbitrary element of the Lie algebra $\mathscr{g}$ of $\mathscr{G}$, i.e.,

$$\left.\frac{dH_A{}^B}{d\tau}\right|_0 \in \mathscr{g}. \tag{28.12}$$

Comparing the conditions (28.11) and (28.12) with the condition (27.33) for a material connection, we see that

$$N^{ij}{}_k{}^A\left(\frac{\partial x^l}{\partial X^E} K_F{}^E\right) \frac{\partial x^k}{\partial X^C} \left(\frac{\partial K_A{}^C}{\partial X^D} + K_A{}^G \Gamma^C_{GD}\right) = 0. \tag{28.13}$$

Consequently, (28.7) may be rewritten as

$$N^{ij}{}_k{}^A\left(\frac{\partial x^l}{\partial X^E} K_F{}^E\right) K_A{}^C \frac{\partial X^D}{\partial x^j} \left(\frac{\partial^2 x^k}{\partial X^C \partial X^D} - \frac{\partial x^k}{\partial X^B} \Gamma^B_{CD}\right) + \varrho b^i = \varrho \frac{\partial^2 x^i}{\partial t^2}. \tag{28.14}$$

Now we claim that the leading coefficient involving the smooth field **K** may be extended from the domain $\mathscr{N}$ to the entire reference configuration $\boldsymbol{\varkappa}(\mathscr{B})$. To prove this fact, we observe the identity

$$N^{ij}{}_k{}^A\left(\frac{\partial x^l}{\partial X^E} K_J{}^E H_F{}^J\right) K_L{}^C H_A{}^L = N^{ij}{}_k{}^A\left(\frac{\partial x^l}{\partial X^E} K_F{}^E\right) K_A{}^C \tag{28.15}$$

for all smooth fields **H** on $\mathscr{N}$ with values belonging to $\mathscr{G}$. Indeed, if the identity (28.15) is true, then the leading coefficient of (28.14) forms a certain globally smooth field, which may be calculated by using any smooth field **K** defined on the neighborhoods $\mathscr{N}$ belonging to the covering of $\boldsymbol{\varkappa}(\mathscr{B})$. Hence (28.14) has the global form

$$G^{ij}{}_k{}^C\left(\frac{\partial x^l}{\partial X^E}, X^A\right) \frac{\partial X^D}{\partial x^j} \left(\frac{\partial^2 x^k}{\partial X^C \partial X^D} - \frac{\partial x^k}{\partial X^B} \Gamma^B_{CD}\right) + \varrho b^i = \varrho \frac{\partial^2 x^i}{\partial t^2}, \tag{28.16}$$

where the globally smooth function $G^{ij}{}_k{}^C(\mathbf{F}, \mathbf{X})$ is defined by the local formula

$$G^{ij}{}_k{}^A(\mathbf{F}, \mathbf{X}) = N^{ij}{}_k{}^A(\mathbf{FK}(\mathbf{X})) K_A{}^C(\mathbf{X}), \tag{28.17}$$

where **K** is any smooth field satisfying (26.2) with its domain containing the point **X**. Since there is such a neighborhood $\mathscr{N}$ and such a smooth field **K** on $\mathscr{N}$ for each point $\mathbf{X} \in \boldsymbol{\varkappa}(\mathscr{B})$, the field $G^{ij}{}_k{}^C(\mathbf{F}, \mathbf{X})$ is well defined by (28.17).

To prove the identity (28.15), we eliminate first the nonsingular factor $K_A{}^C$, obtaining

$$N^{ij}{}_k{}^D\left(\frac{\partial x^l}{\partial X^E} K_B{}^E H_C{}^B\right) H_D{}^A = N^{ij}{}_k{}^A\left(\frac{\partial x^l}{\partial X^E} K_C{}^E\right), \tag{28.18}$$

which must hold for all smooth fields $[H_D{}^A]$ with values in $\mathscr{G}$. We can verify this identity by using the symmetry condition (28.9), which may be written as

$$N^{ij}(F_B{}^k H_A{}^B) = N^{ij}(F_A{}^k) \tag{28.19}$$

for all $[H_A{}^B] \in \mathscr{G}$ and for all deformation gradients $[F_A{}^k]$. Differentiating (28.19) with respect to $F_A{}^k$ and using the definition (28.8), we get

$$N^{ij}{}_k{}^D(F_B{}^l H_C{}^B) H_D{}^A = N^{ij}{}_k{}^A(F_C{}^l), \tag{28.20}$$

which implies (28.18) directly by setting

$$F_C{}^l = \frac{\partial x^l}{\partial X^E} K_C{}^E. \tag{28.21}$$

Thus the global form (28.16) is established.

As before the matrix $[\partial X^D/\partial x^i]$ may be absorbed into the leading coefficient $G^{ij}{}_k{}^C$. Then we can rewrite (28.16) as

$$H^{iD}{}_k{}^C\left(\frac{\partial x^l}{\partial X^E}, X^A\right)\left(\frac{\partial^2 x^k}{\partial X^C\,\partial X^D} - \frac{\partial x^k}{\partial X^B}\,\Gamma^B_{CD}\right) + \varrho_\varkappa b^i = \varrho_\varkappa \frac{\partial^2 x^i}{\partial t^2}, \tag{28.22}$$

where

$$H^{iD}{}_k{}^C(\mathbf{F}, \mathbf{X}) \equiv (\det \mathbf{F})\mathbf{G}^{ij}{}_k{}^C(\mathbf{F}, \mathbf{X})(F^{-1})_j{}^D. \tag{28.23}$$

The system (28.22) governs the deformation functions $x^i(X^A, t)$, $i = 1, 2, 3$, relative to the reference configuration $\boldsymbol{\varkappa}(\mathscr{B})$.

It should be noted that the combination involving the Γ symbols of a material connection on the left-hand side of the equations of motion (28.22) corresponds precisely to the covariant derivative of the deformation gradient $F_C{}^k$ taken relative to the material connection, viz.,

$$F^k_{C,D} = \frac{\partial F_C{}^k}{\partial X^D} - F_B{}^k \Gamma^B_{CD}, \tag{28.24}$$

where

$$F_C{}^k \equiv \frac{\partial x^k}{\partial X^C}. \tag{28.25}$$

It is not surprising that the covariant derivative of $\mathbf{F}$ with respect to a material connection is important for the equations of motion, since the parallel transports of a material connection are all material isomorphisms, which leave the response function invariant.

It should be noted that the covariant derivative $F^k_{C,D}$ given by (28.24) cannot vanish when the body is not locally homogeneous. In fact the conditions of integrability for the second-order partial differential equations

$$\frac{\partial^2 x^k}{\partial X^C\, \partial X^D} - \frac{\partial x^k}{\partial X^B}\, \Gamma^B_{CD} = 0 \tag{28.26}$$

are precisely the vanishing of both the torsion tensor $\boldsymbol{\Upsilon}$ and the curvature tensor $\mathbf{R}$ for the material connection. When these conditions are satisfied, we can find local deformation functions $\bar{X}^B = \bar{X}^B(X^A)$ relative to which $\bar{\Gamma}^A_{BC}$ vanish identically. In other words, the Euclidean connection in the configuration $\bar{\boldsymbol{\varkappa}}$ with coordinates $(\bar{X}^A)$ is a material connection. Hence $\bar{\boldsymbol{\varkappa}}$ is a homogeneous configuration.

In the special case when the body is globally homogeneous, we choose $\boldsymbol{\varkappa}$ to be a homogeneous reference configuration. Then a material connection on $\boldsymbol{\varkappa}(\mathscr{B})$ is just the Euclidean connection with Γ^A_{BC} vanishing identically, and the system (28.22) reduces to the previous system (28.4).

The equations of motion (28.22) were derived originally by Noll.(21) In the following section we shall obtain some static and dynamic universal solutions for certain inhomogeneous incompressible isotropic elastic bodies, known as *laminated bodies*.

29. Inhomogeneous Isotropic Elastic Solid Bodies

In this section we apply the theory of materially uniform smooth elastic body to an isotropic solid. As before we choose a reference configuration $\boldsymbol{\varkappa}(\mathscr{B})$ for the body, and we assume that the response function $\mathbf{G}$ has the representation (26.2). The assumption that the material is an isotropic solid means that we can choose the nominal response function $\mathbf{N}(\mathbf{F})$ to have a symmetry group $\mathscr{G} = \mathscr{SO}(\mathscr{V})$. That is to say, $\mathbf{N}(\mathbf{F})$ satisfies the condition

$$\mathbf{N}(\mathbf{FQ}) = \mathbf{N}(\mathbf{F}) \tag{29.1}$$

for all rotations $\mathbf{Q}$. As we have explained in Section 19, a general solution

(21) N. Noll, Materially uniform simple bodies with inhomogeneities, *Archive for Rational Mechanics and Analysis*, Vol. 27, pp. 1–32, 1967. The derivation summarized here is given by Wang, C.-C., On the geometric structures of simple bodies, a mathematical foundation for the theory of continuous distributions of dislocations, *Archive for Rational Mechanics and Analysis*, Vol. 27, pp. 33–94, 1967.

for this condition is

$$\mathbf{N}(\mathbf{F}) = \bar{\mathbf{N}}(\mathbf{B}), \tag{29.2}$$

where **B** denotes the left Cauchy–Green tensor, viz.,

$$\mathbf{B} = \mathbf{F}\mathbf{F}^T. \tag{29.3}$$

The principle of material frame-indifference implies that $\bar{\mathbf{N}}$ is an isotropic function; i.e.,

$$\bar{\mathbf{N}}(\mathbf{Q}\mathbf{B}\mathbf{Q}^T) = \mathbf{Q}\bar{\mathbf{N}}(\mathbf{B})\mathbf{Q}^T \tag{29.4}$$

for all orthogonal tensors **Q**.

In Section 26 we remarked that the field **K** in the representation (26.2) is unique to within an arbitrary right multiplication by a field **H** with values belonging to the nominal symmetry group $\mathscr{G}$. In particular, if the body is an isotropic solid, then any two fields **K** and $\bar{\mathbf{K}}$ in (26.2) differ from each other by a rotation **Q**:

$$\bar{\mathbf{K}} = \mathbf{K}\mathbf{Q}. \tag{29.5}$$

A necessary and sufficient condition for (29.5) is

$$\bar{\mathbf{K}}\bar{\mathbf{K}}^T = \mathbf{K}\mathbf{K}^T \tag{29.6}$$

or, equivalently,

$$(\bar{\mathbf{K}}^{-1})^T\bar{\mathbf{K}}^{-1} = (\mathbf{K}^{-1})^T\mathbf{K}^{-1}. \tag{29.7}$$

Hence we may define a metric **M** on $\varkappa(\mathscr{B})$ by

$$\mathbf{M} = (\mathbf{K}^{-1})^T\mathbf{K}^{-1}, \qquad \mathbf{M}^{-1} = \mathbf{K}\mathbf{K}^T. \tag{29.8}$$

Since for each point $\mathbf{X}_0 \in \varkappa(\mathscr{B})$ there is a neighborhood $\mathscr{N}_0$ on which a smooth field **K** satisfying the representation (26.2) exists, the metric **M** defined by (29.8) is smooth; i.e., **M** is a Riemannian metric on $\varkappa(\mathscr{B})$.

We call **M** the *characteristic metric* on $\varkappa(\mathscr{B})$, since it characterizes the fields **K** completely by the condition (29.8). It follows that all material isomorphisms among points in $\varkappa(\mathscr{B})$ are also determined by **M**. In fact an isomorphism $\mathbf{K}(\mathbf{X}, \mathbf{Y}): \mathscr{V}(\mathbf{X}) \to \mathscr{V}(\mathbf{Y})$ is a material isomorphism if and only if it preserves the metric **M** in the sense that

$$\mathbf{u} \cdot \mathbf{M}(\mathbf{X})\mathbf{w} = \mathbf{K}(\mathbf{X}, \mathbf{Y})\mathbf{u} \cdot \mathbf{M}(\mathbf{Y})\mathbf{K}(\mathbf{X}, \mathbf{Y})\mathbf{w} \tag{29.9}$$

for all vectors **u** and **w** in $\mathscr{V}(\mathbf{X})$. To prove this fact, we recall first that a material isomorphism $\mathbf{K}(\mathbf{X}, \mathbf{Y})$ may be represented by (27.6). Using it and the definition (29.8), we see that the left-hand side of (29.9) is equal to $\mathbf{K}^{-1}(\mathbf{X})\mathbf{u} \cdot \mathbf{K}^{-1}(\mathbf{X})\mathbf{w}$ while the right-hand side is given by

$$\begin{aligned}\mathbf{K}(\mathbf{Y})\mathbf{Q}\mathbf{K}^{-1}(\mathbf{X})\mathbf{u} &\cdot \mathbf{K}^{-1}(\mathbf{Y})^T\mathbf{K}^{-1}(\mathbf{Y})\mathbf{K}(\mathbf{Y})\mathbf{Q}\mathbf{K}^{-1}(\mathbf{X})\mathbf{w} \\ &= \mathbf{Q}\mathbf{K}^{-1}(\mathbf{X})\mathbf{u} \cdot \mathbf{Q}\mathbf{K}^{-1}(\mathbf{X})\mathbf{w} = \mathbf{K}^{-1}(\mathbf{X})\mathbf{u} \cdot \mathbf{K}^{-1}(\mathbf{X})\mathbf{w}. \qquad (29.10)\end{aligned}$$

Conversely, if we write $\mathbf{K}(\mathbf{X}, \mathbf{Y})$ in the form (27.6), then from (29.9) we obtain

$$\mathbf{K}^{-1}(\mathbf{X})\mathbf{u} \cdot \mathbf{K}^{-1}(\mathbf{X})\mathbf{w} = \mathbf{H}\mathbf{K}^{-1}(\mathbf{X})\mathbf{u} \cdot \mathbf{H}\mathbf{K}^{-1}(\mathbf{X})\mathbf{w}. \qquad (29.11)$$

Thus $\mathbf{H} \in \mathscr{G}$, and $\mathbf{K}(\mathbf{X}, \mathbf{Y})$ is a material isomorphism.

The criterion (29.9) for a material isomorphism $\mathbf{K}(\mathbf{X}, \mathbf{Y})$ is a convenient property for the determination of a material connection. In Section 56, IVT-2, we have explained that the Riemannian connection possesses the basic property that the Riemannian metric is preserved by the parallel transports along any smooth curve. By virtue of (29.9) this property corresponds precisely to the requirement of a material connection when the metric is the characteristic metric $\mathbf{M}$. As a result, the Riemannian connection associated with $\mathbf{M}$ is a material connection on $\boldsymbol{\varkappa}(\mathscr{B})$ for an isotropic solid body $\mathscr{B}$. In particular, the Christoffel symbols of $\mathbf{M}$ are the Γ symbols of the material connection.

Note. The Riemannian connection is the unique affine connection which possesses the following two properties: (i) the Riemannian metric is a covariantly constant field, and (ii) the torsion tensor vanishes. The preceding assertion is a well-known result in classical differential geometry. To prove that result, we use the formula

$$M_{AB,C} = \frac{\partial M_{AB}}{\partial X^C} - M_{DB}\Gamma^D_{AC} - M_{AD}\Gamma^D_{BC} \qquad (29.12)$$

for the covariant derivative of a metric $\mathbf{M}$ with respect to an affine connection in general. Then the condition (i) implies that

$$\frac{\partial M_{AB}}{\partial X^C} = M_{DB}\Gamma^D_{AC} + M_{AD}\Gamma^D_{BC}, \qquad (29.13)$$

and the condition (ii) implies that

$$\Gamma^A_{BC} = \Gamma^A_{CB}. \qquad (29.14)$$

Now we rotate the indices A, B, C in (29.13), obtaining

$$\frac{\partial M_{BC}}{\partial X^A} = M_{DC}\Gamma^D_{BA} + M_{BD}\Gamma^D_{CA}, \tag{29.15}$$

$$\frac{\partial M_{CA}}{\partial X^B} = M_{DA}\Gamma^D_{CB} + M_{CD}\Gamma^D_{AB}, \tag{29.16}$$

Adding (29.13) with (29.15) and subtracting (29.16), then using the symmetry conditions of the connection and the metric, we obtain

$$\frac{\partial M_{AB}}{\partial X^C} + \frac{\partial M_{BC}}{\partial X^A} - \frac{\partial M_{CA}}{\partial X^B} = 2M_{BD}\Gamma^D_{CA}. \tag{29.17}$$

Thus the Γ symbols must be the Christoffel symbols, viz.,

$$\Gamma^D_{CA} = \frac{1}{2} M^{BD}\left(\frac{\partial M_{AB}}{\partial X^C} + \frac{\partial M_{BC}}{\partial X^A} - \frac{\partial M_{CA}}{\partial X^B}\right) = \begin{Bmatrix} D \\ CA \end{Bmatrix}. \tag{29.18}$$

The preceding result shows that, for an isotropic solid body $\mathscr{B}$ the geometric structure associated with the collection of fields **K** is just the Riemannian structure induced by the characteristic metric **M**. In component form relative to (X^A) **M** is given by

$$M_{AB} = \delta_{CD}(K^{-1})_A{}^C(K^{-1})_B{}^D, \qquad M^{AB} = \delta^{CD}K_C{}^A K_D{}^B. \tag{29.19}$$

In general M_{AB} and M^{AB} are certain smooth functions of the coordinates (X^D). In Section 59, IVT-2, we explained that a necessary and sufficient condition for **M** to be a flat or locally Euclidean metric is the vanishing of the Riemann–Christoffel curvature tensor **R**. In the context of isotropic solids the preceding result means that $\mathscr{B}$ is locally homogeneous if and only if **R** vanishes. That is to say, $\mathbf{R} = \mathbf{0}$ if and only if for each $\mathbf{X}_0 \in \boldsymbol{\varkappa}(\mathscr{B})$ there is a neighborhood $\mathscr{N}_0$ such that we can find certain deformation functions $\bar{X}^A = \bar{X}^A(X^C)$, $A = 1, 2, 3$, which transform M_{AB} into δ_{AB}, viz.,

$$\delta_{AB} = M_{CD}\frac{\partial X^C}{\partial \bar{X}^A}\frac{\partial X^D}{\partial \bar{X}^B}, \qquad \delta^{AB} = M^{CD}\frac{\partial \bar{X}^A}{\partial X^C}\frac{\partial \bar{X}^B}{\partial X^D}. \tag{29.20}$$

Note. If the deformation functions are defined for all points in $\boldsymbol{\varkappa}(\mathscr{B})$, and if (29.20) holds, then the body is globally homogeneous. However, $\mathbf{R} = \mathbf{0}$ is only necessary but not sufficient for **M** to be globally Euclidean (i.e., not sufficient for $\mathscr{B}$ to be globally homogeneous). By virtue of the preceding remark the field **R** for the characteristic metric **M** may be called the *local inhomogeneity* of $\mathscr{B}$ in the reference configuration $\boldsymbol{\varkappa}(\mathscr{B})$.

Now suppose that the characteristic metric $\mathbf{M}$ on $\varkappa(\mathscr{B})$ is given explicitly. Then the stress tensor field in any deformed configuration $\boldsymbol{\chi}(\mathscr{B})$ may be determined in the following way: First, according to the representation (26.2),

$$\mathbf{T} = \mathbf{N}(\mathbf{FK}). \tag{29.21}$$

But from (29.2) the right-hand side of (29.21) has the representation

$$\mathbf{T} = \bar{\mathbf{N}}(\mathbf{FKK}^T\mathbf{F}^T). \tag{29.22}$$

From (29.8) the components of the argument of $\bar{\mathbf{N}}$ are

$$(FKK^TF^T)^{kl} = m^{kl} = M^{AB}\frac{\partial x^k}{\partial X^A}\frac{\partial x^l}{\partial X^B}. \tag{29.23}$$

Then from (29.22) the components of the stress tensor $\mathbf{T}$ are

$$T^{ij} = \bar{\mathbf{N}}^{ij}(m^{kl}), \tag{29.24}$$

where the function $\bar{\mathbf{N}}$ is isotropic; cf. (29.4).

If the isotropic solid is incompressible, then as before the deformation must be isochoric, and the stress tensor is determined by the deformation to within an arbitrary additive pressure only. Thus the constitutive equation has the representation

$$\mathbf{T} = -p\mathbf{I} + \bar{\mathbf{N}}(\mathbf{FKK}^T\mathbf{F}^T), \tag{29.25}$$

where the determinate stress $\mathbf{S} = \mathbf{T} + p\mathbf{I} = \bar{\mathbf{N}}(\mathbf{FKK}^T\mathbf{F}^T)$ is required to satisfy the condition

$$\operatorname{tr}\mathbf{S} = 0. \tag{29.26}$$

We shall now consider a special class of incompressible isotropic solid bodies, called *laminated bodies*. Certain families of static and dynamic universal solutions for this class of bodies may be obtained directly by using the representation (29.25).

First, we call $\mathscr{B}$ a *laminated plate* if relative to a Cartesian coordinate system (X, Y, Z) in a certain reference configuration $\varkappa(\mathscr{B})$ the component matrix of the characteristic metric is of the form

$$[M^{AB}] = \begin{bmatrix} M^{11} & 0 & 0 \\ 0 & M^{22} & M^{23} \\ 0 & M^{23} & M^{33} \end{bmatrix}, \tag{29.27}$$

where the nonzero components M^{11}, M^{22}, M^{33}, and M^{23} are functions of X only. The inverse matrix $[M_{AB}] = [M^{AB}]^{-1}$ has a similar form, and the nonzero components M_{11}, M_{22}, M_{33}, and M_{23} are also functions of X only. We claim that the following two families of deformations are static universal solutions:

Family 0. The deformation functions are

$$x = AX, \qquad y = BY + CZ, \qquad z = DY + EZ, \tag{29.28}$$

where $A, \ldots, E$ are constants such that

$$A(BE - CD) = 1. \tag{29.29}$$

The deformations in this family are homogeneous. In the deformed configuration the component matrix $[m^{kl}]$ is of the form

$$[m^{kl}] = \begin{bmatrix} m^{11} & 0 & 0 \\ 0 & m^{22} & m^{23} \\ 0 & m^{23} & m^{33} \end{bmatrix}, \tag{29.30}$$

where the nonzero components m^{11}, m^{22}, m^{33}, and m^{23} are functions of x only. Indeed, from (29.23), (29.27), and (29.28)

$$\begin{aligned} m^{11}(x) &= A^2 M^{11}\left(\frac{x}{A}\right), \\ m^{22}(x) &= B^2 M^{22}\left(\frac{x}{A}\right) + 2BCM^{23}\left(\frac{x}{A}\right) + C^2 M^{33}\left(\frac{x}{A}\right), \\ m^{33}(x) &= D^2 M^{22}\left(\frac{x}{A}\right) + 2DEM^{23}\left(\frac{x}{A}\right) + E^2 M^{33}\left(\frac{x}{A}\right), \\ m^{23}(x) &= BDM^{22}\left(\frac{x}{A}\right) + (CD + BE)M^{23}\left(\frac{x}{A}\right) + CEM^{33}\left(\frac{x}{A}\right). \end{aligned} \tag{29.31}$$

Now since the determinate stress $\mathbf{S}$ is given by an isotropic function of $\mathbf{m}$, its components in (x, y, z) must satisfy the following conditions: (a) $S_{\langle xy \rangle} = S_{\langle xz \rangle} = 0$ and (b) $S_{\langle xx \rangle}$, $S_{\langle yy \rangle}$, $S_{\langle zz \rangle}$, and $S_{\langle yz \rangle}$ depend only on x. In Section 25 we explained that the equations of equilibrium are integrable in this case with the pressure field given by (25.12) and the stress components given by (25.13).

Family 1. The deformation functions are

$$r = (2AX)^{1/2}, \qquad \theta = BY + CZ, \qquad z = DY + EZ, \tag{29.32}$$

where $A, \ldots, E$ are constants satisfying the condition (29.29). For this family the coordinate system (r, θ, z) in the deformed configuration is the cylindrical system.

We can prove that the deformations in this family are static universal solutions by using an argument similar to that of the preceding family. Indeed, it can be shown that the physical components of the determinate stress satisfy the conditions (a) $S_{\langle r\theta\rangle} = S_{\langle rz\rangle} = 0$ and (b) $S_{\langle rr\rangle}$, $S_{\langle\theta\theta\rangle}$, $S_{\langle zz\rangle}$, and $S_{\langle\theta z\rangle}$ depend only on r. Thus the equations of equilibrium (25.14) are integrable with the pressure field given by (25.15) and the physical components of the stress tensor field given by (25.16).

Next, we call $\mathscr{B}$ a *laminated cylindrical shell* if relative to a cylindrical coordinate system (R, Θ, Z) in a certain reference configuration $\boldsymbol{\varkappa}(\mathscr{B})$ the component matrix of the characteristic metric is of the form (29.27), where the nonzero components M^{11}, M^{22}, M^{33}, and M^{23} are functions of R only. Now we claim that the following two families of deformations are static universal solutions:

Family 2. The deformation functions are

$$x = \frac{1}{2} AR^2, \qquad y = B\Theta + CZ, \qquad z = D\Theta + EZ, \tag{29.33}$$

where $A, \ldots, E$ satisfy (29.28).

Family 3. The deformation functions are

$$r = (AR^2 + B)^{1/2}, \qquad \theta = C\Theta + DZ, \qquad z = D\Theta + EZ, \tag{29.34}$$

where $A, \ldots, F$ are constants satisfying the condition

$$A(CF - DE) = 1. \tag{29.35}$$

The proof that these families are static universal solutions is essentially the same as before. Specifically, for Family 2 the determinate stress in the deformed configuration satisfies the conditions of the planar problem in Section 25, while for Family 3 the determinate stress satisfies the conditions for the first type of the cylindrical problems in the same section. Hence regardless of what incompressible isotropic elastic solid material the laminated cylindrical shell is made up of, the equations of equilibrium are always integrable.

Finally, we call $\mathscr{B}$ a *laminated spherical shell* if relative to a spherical coordinate system (R, Θ, Φ) in a certain reference configuration $\boldsymbol{\varkappa}(\mathscr{B})$ the physical components of the characteristic metric $\mathbf{M}$ satisfy the following conditions:

$$M_{\langle R\Theta\rangle} = M_{\langle R\Phi\rangle} = M_{\langle \Theta\Phi\rangle} = 0, \tag{29.36}$$

$$M_{\langle \Theta\Theta\rangle} = M_{\langle \Phi\Phi\rangle}, \tag{29.37}$$

and the nonzero components $M_{\langle RR\rangle}$, $M_{\langle \Theta\Theta\rangle} = M_{\langle \Phi\Phi\rangle}$ are functions of R only. In this case we claim that the following family of deformations are static universal solutions:

Family 4. The deformation functions are

$$r = (\pm R^3 + A)^{1/3}, \qquad \theta = \pm\Theta, \qquad \varphi = \Phi, \tag{29.38}$$

where A is an arbitrary constant.

For this family the determinate stress in the deformed configuration satisfies the conditions of the spherical problems considered in Section 25. Thus the equations of equilibrium are integrable with the pressure field given by (25.21) and the stress components given by (25.22).

As we explained in Section 25, if we replace the constants in the preceding families of static universal solutions by appropriate functions of t in such a way that the resulting motions have a conservative acceleration field, then we obtain families of dynamic universal solutions. Since the conditions on the functions of t are independent of the characteristic metric, the results are exactly the same as those of the homogeneous bodies.

For more details on the universal solutions of laminated bodies, we refer the reader to the original paper by Wang.[22]

[22] C.-C. Wang, Universal solutions for incompressible laminated bodies, *Archive for Rational Mechanics and Analysis*, Vol. 29, pp. 161–192, 1968. Reprinted in W. Noll, R. A. Toupin, and C.-C. Wang, *Continuum Theory of Inhomogeneities in Simple Bodies*, Springer, Berlin, 1968.

Selected Reading for Part A

ABRAHAM, R., *Foundations of Mechanics*, Benjamin, New York, 1967.

AMES, J. S., and MURNAGHAM, F. D., *Theoretical Mechanics*, Ginn, Boston, 1929, Dover Publications, New York, 1958.

BOWEN, R. M., and WANG, C.-C., *Introduction to Vectors and Tensors*, Volumes 1 and 2, Plenum, New York, 1976.

COLEMAN, B. D., MARKOVITZ, H., and NOLL, W., *Viscometric Flows of Non-Newtonian Fluids*, Springer, Berlin, 1966.

FLANDERS, H., *Differential Forms*, Academic Press, New York, 1963.

GOLDSTEIN, H., *Classical Mechanics*, Addison-Wesley, Cambridge, Massachusetts, 1950.

LANDAU, L. D., and LIFSHITZ, E. M., *Mechanics*, Course of Theoretical Physics, Volume 1, Addison-Wesley, Reading, Massachusetts, 1960.

NOLL, W., TOUPIN, R. A., and WANG, C.-C., *Continuum Theory of Inhomogeneities in Simple Bodies*, Springer, Berlin, 1968.

SOMMERFELD, A. J. W., *Mechanics*, Lectures on Theoretical Physics, Volume 1, Academic Press, New York, 1952.

STERNBERG, S., *Lectures on Differential Geometry*, Prentice-Hall, Englewood Cliffs, New Jersey, 1964.

SYNGE, J. L., Classical dynamics, in Flügge's *Handbuch der Physik*, Band III/1, Springer, Berlin, 1960.

TRUESDELL, C., *A First Course in Rational Continuum Mechanics*, Academic Press, New York, 1977.

TRUESDELL, C., Editor, *Continuum Mechanics*, International Science Review Series, Volume VIII, Parts 1–4, Gordon and Breach, New York, 1966.

TRUESDELL, C., and NOLL, W., The non-linear field theories of mechanics, in Flügge's *Handbuch der Physik*, Band III/3, Springer, Berlin, 1965.

TRUESDELL, C., and TOUPIN, R. A., The classical field theories, in Flügge's *Handbuch der Physik*, Band III/1, Springer, Berlin, 1960.

WANG, C.-C., and TRUESDELL, C., *Introduction to Rational Elasticity*, Noordhoff, Leyden, 1973.

WHITTAKER, E. T., *A Treatise on the Analytical Dynamics of Particles and Rigid Bodies*, Cambridge University Press, Cambridge, England, 1964.

Index

Pages 1–198 will be found in Part A, pages 199–386 in Part B.